Problems of Scientific Revolution

Problems of Scientific Revolution

Progress and obstacles to progress in the sciences

THE HERBERT SPENCER LECTURES 1973

EDITED BY
ROM HARRÉ

9783

CLARENDON PRESS · OXFORD
1975

Oxford University Press, Ely House, London W.1

GLASGOW NEW YORK TORONTO MELBOURNE WELLINGTON
CAPE TOWN IBADAN NAIROBI DAR ES SALAAM LUSAKA ADDIS ABABA
DELHI BOMBAY CALCUTTA MADRAS KARACHI LAHORE DACCA
KUALA LUMPUR SINGAPORE HONG KONG TOKYO

CASEBOUND ISBN 0 19 58211 0
PAPERBACK ISBN 0 19 58213 7

© OXFORD UNIVERSITY PRESS 1975

PRINTED IN GREAT BRITAIN BY
RICHARD CLAY (THE CHAUCER PRESS) LTD
BUNGAY, SUFFOLK

Preface

THE Herbert Spencer lectures were founded seventy years ago, by the generous benefaction of S. Krishnavarma. Until 1970 Herbert Spencer was commemorated by an annual lecture. In that year a series of lectures was given, published in 1972 as *Biology and the human sciences*, edited by J. W. S. Pringle (Clarendon Press, Oxford). The success of the series encouraged the Board of Management to repeat the experiment, with very gratifying results, not only in the quality of the lectures but in the size and enthusiasm of the audiences. For the 1973 series the Board of Management conceived the idea of asking six distinguished lecturers, including scientists, a historian of science, and a philosopher, to examine the notion of progress in science, and to reflect on that progress, so central to Spencer's thought, and so dear to his political and moral convictions, both with respect to its consequences and to the obstacles it might encounter. The lecturers were asked to take as their general title *Progress and obstacles to progress in the sciences*.

Spencer had supposed that while evolution was a necessary process, obeying his famous principle that development is from 'relatively indefinite incoherent homogeneity to relatively definite coherent heterogeneity' it proceeded to no particular end, and those structures which were produced would, by the action of the same forces, ultimately be dissolved. The process of evolution was supposed to characterize every product. Spencer claimed that 'not only is the law exemplified in the evolution of the social organism, but it is exemplified in all products of human thought and action, whether concrete or abstract, real or ideal' (*First principles*, p. 279). Though the dissolution of such products was as inevitable as their production, he supposed that great heights of stable civilization would be reached between the developmental and destructive phases of an evolutionary process.

Though the reader will soon sense that for none of the lecturers is Spencer's qualified optimism wholly dead, each in his own way draws our attention both to the complexities of the way scientific knowledge grows, and to the equivocal and uncertain character of many of the possibilities that growth opens up. Even the much qualified and elaborated idea of scientific progress that emerges from the lectures assembled in this series is far from being a guide to inevitable progress in the moral and political life of mankind. But one may be sure that Spencer would have been pleased to discover the central place the evolutionary analogy plays in much of the philosophical discussions of the notion of progress to be found in this volume.

R.H.

Linacre College, Oxford
1974

Contents

1 *What is progress in science?*

SIR HERMANN BONDI

Ministry of Defence, London

I HAVE the task of starting a series where the title can be taken in many different ways. Obstacles to progress may be within the science itself or outside. You can look at the problem psychologically, politically, or sociologically. I, myself, in choosing to talk about *progress*, have, I think, a slightly easier task than those who will talk of its obstacles, because progress in science is presumably something that can be defined in internal terms intrinsic to science. My thesis will be that the concept of progress in science is not a simple concept. You might well say that it does not need a long lecture to explain that something is *not* simple. Nevertheless, I will try to dispel any prejudices that it might be simple.

To start then, I think I should try and define what science is. This is a subject in which I am a whole-hearted follower of Karl Popper and his criterion of demarcation, which is why I feel so honoured to have been asked to start a series which will be completed by Karl Popper himself, for whom I have such admiration.

I want to describe what Popper's concept is, in a very compressed form. In his view it is the task of the scientist to propose a theory, a theory that must naturally encompass what is known at the time, but that over and above this *must* make forecasts of what future experiments or observations will show. If such experiments are then performed and the outcome is in agreement with what the theory has said, then we must *never* say that the theory has been proved. We can only say that it has stood a test successfully and it is the task of the theory to go on making further deductions, further statements that can be tested empirically. If, however, such a test goes against the theory, then the theory has been disproved and then one has got to start with a new theory which may be a very difficult matter. This primacy of disproof is an essential part of the scheme, and it grows naturally from the logical structure: a scientific theory is always a *general* statement that all things of this kind behave in a certain way. To put it in the most elementary way of which I am capable, you *may presume* from observation that *any* apples, when released, will drop, but you cannot possibly *deduce* this rule from your observations, because the theory that any apple when released will drop applies to all apples, past, present, and future. You may have tried it out with ten, or with a hundred, or with a thousand apples, but you will certainly not have tried it out with *all* apples, and therefore you can never prove the theory. But find only *one* apple that, when released, rises, and you have disproved the theory. With a particular instance you can disprove the general statement, but you can never logically deduce a general statement from a particular, however large, sample of experiments. The fact that you cannot deduce your scientific theory by any logical method shows that the vital part of the subject is originality and imagination. Formulating a scientific theory always involves an imaginative leap. We cannot imagine a mechanical process for going from a set of experiments or observations to a theory. A theory will be scientific if and only if it can be empirically disproved. I think this is a very important point. I suffer more at some times in my life than others from getting large numbers of letters with brilliant scientific ideas. The largest proportion of them, of course, comes from cranks. But with clever cranks it is not always immediately obvious that the writer is a crank and not somebody who has got perhaps quite a good idea oddly expressed. But when I come to the line, to which one often comes, 'Well, if the experiments go this way, they will fit in with my theory, and if they go any other way they will also fit in with my theory, so you see how good a theory it is', then my ample waste-paper basket is the right destination.

It is immediately clear, then, that with disproof playing such a very important role, we cannot expect any description of the progress of science to be smooth and straight. Indeed, it must be something very far from it. Let me again give you an example, perhaps the most extreme example that we have got, of Popper's criteria in action. When Newton's theory of gravitation was proposed, over three hundred years ago, there followed enormous numbers of tests.

The theory was used by astronomers to predict the position of planets and satellites and to forecast when eclipses would occur, then to go backwards in time and date historical events by them—they would refer historical events to eclipses by calculating exactly when an eclipse would occur. To say that that theory was tested a hundred thousand times and passed each test brilliantly is, I am sure, numerically an understatement. But the fact that it had been tested so very very often did not help the theory all that much when the first discrepancies between theory and observation emerged towards the end of the last century, with more and more accurate observations of the motion of the planets and more and more accurate methods of calculating the mutual perturbation of the planets. Newcombe then discovered before the turn of the century that there was a discrepancy in the motion of the planet Mercury (he referred to this in his calculations as the 'empirical term'). With hindsight, we can say that to a large extent this was a disproof of Newton's theory of gravitation, although in fact the progress to Einstein's theory came from a rather different side, a change in the intellectual climate, but the disproof was really there with Newcombe. And so you have the result that a theory that had been tested exhaustively, and passed all those tests brilliantly, was still liable to be disproved.

This disproof had a tremendous effect on the intellectual climate. It had been thought that whatever in the world might be difficult, might be complex, might be hard to understand, at least Newton's theory of gravitation was good and solid, tested well over a hundred thousand times. And when such a theory falls victim to the increasing precision of observation and calculation, one certainly feels that one can never again rest assured. This is the stuff of progress. You cannot therefore speak of progress as progress in a particular direction, as a progress in which knowledge becomes more and more certain and more and more all-embracing. At times we make discoveries that sharply reduce the knowledge that we have, and it is discoveries of this kind that are indeed the seminal point in science. It is they that are the real roots of progress and lead to the jumps in understanding, but in the first instance they reduce what we regard as assured knowledge.

It is, of course, important to remember that when a theory has passed a very large number of tests, like Newton's theory, and is then disproved—and we can certainly speak of its disproof now—you would not say that everything that was tested before—all those forecasts—were wrong. They were right, and you know therefore that although the theory *qua* general theory is no longer tenable, yet it is something that described a significant volume of experience quite well. And indeed, although we have a newer and better theory of gravitation—Einstein's theory and one or two variants of it in addition—nevertheless, whenever we do not want to carry out calculations of the motions of planets and satellites with extreme precision—we use Newton's theory because it is simpler. The difference lies in our attitude to it. When you formulate a theory it has universalist totalitarian claims. When you have tested it, then you know that that theory describes certain ranges of phenomena, a certain *limited* range, with a certain limited measure of accuracy and precision. And that description will hold. It is your interpretation of the theory as something that is the Truth (with a capital T) (a term that to my thinking has nothing to do with science anyway), that will have been shattered. And you will naturally look for a description that is *better* in the sense that it embraces a wider range of phenomena to higher precision, but is also disprovable. If it is not disprovable it is not scientifically interesting. So, the nature of progress is that crucial points in it are the *shattering* of knowledge, rather than the gaining of knowledge. But I would be quite wrong to say that the gaining of knowledge was not also progress. It just is not a very straight path.

It is implicit in Popper's description that we can perform more searching experiments tomorrow than we can perform today, that you can test the theory more thoroughly tomorrow than you can test it today. Otherwise there wouldn't be all that much point in this vulnerability in forecasting. Our faith that tomorrow we can test more searchingly than today is based on our faith in the progress of technology. It is the technology that supplies the experimenter and the observer with their means. It is a progressive technology that allows us to measure new things, or to measure old things more precisely. And what is the basis of the progress of technology? Of course, it is the progress of science; and so the idea that science, all beautiful and shining, leads, and technology follows, is a total misconception. The relation between the two is the relation between the chicken and the egg and progress in the one is often due to progress in the other as Popper has so well stated. There are in science itself, nevertheless, criteria which makes us regard certain things as progress: namely that we can deal with more searching experiments. We can cover a larger area although our method of covering may by no means be simple.

The search for simplicity is a powerful driving force, but it does not necessarily get us there. The search for simplicity belongs more properly to the subject of the psychology of science than the subject of what constitutes progress, and we know from many examples in history that we have arrived at greater simplicity only to go forward to greater complexity afterwards. Some subjects get into states where they are very hard to describe, let alone understand, because they are so highly complex, although not so many years before there was a phase where they were very simple. We know that in certain circumstances we can hardly expect to find simplicity. On our own scale we know there is a lot of complexity because we know *we* are very complex indeed, so one's hope for simplicity is either to go to the very large or to the very small. Sometimes one is rewarded for certain periods in having a simpler situation and you say, 'Isn't it a marvellous piece of progress to know that we now know *all* the particles that go into making up an atom': only a few years later your simplicity is shattered and you discover that there is a whole zoo of particles. No doubt the time will come when they are again ordered. A very pretty story of the kind of zig-zag that is characteristic of progress in science concerns the atomic weights. When they were first determined with low precision it was thought likely that they all bore integral ratios to each other, that they were all whole-number multiples of some particular unit. Then, when it became possible to measure atomic weights more precisely, one discovered that this was certainly not the case. Next the brilliant idea of isotopes came along, viz. that elements might be made up of atoms the same in all respects other than their masses, and that different masses always occurred with the same relative abundance. This could be tested with precision, and again you could say they were all whole numbers. Before long the precision of measurement had increased yet further and one found that even the individual isotopes bore no whole-number ratios to each other. This way one found the mass defects which express the binding energy of the nuclei. Before very long one appreciated again then that the whole-number property was a very important property but it referred to the number of protons and neutrons in the nucleus of each isotope and not to the mass of the nucleus. And so, by somewhat re-defining your terms, you have had a whole chain: whole-number, not whole-number, whole-number, not whole-number; and that is called progress. And undoubtedly in our understanding it was progress, but it is not rectilinear in any sense.

Sometimes you have this zig-zag interaction between theory and experiment which is integral to the whole structure of Popper's criterion of demarcation, sometimes you go exactly in the opposite direction to where you intend to go. If I may tell a personal story of ten or fifteen years ago, Lyttleton and I published a paper suggesting that perhaps the charges of the proton and the electron were not exactly equal and opposite. We found that under certain conditions, with a certain magnitude of that difference, one could come to quite interesting results on the theoretical side both in the structure of the necessary modification of Maxwell's equations and in the structure of the universe (I do not say it was a particularly good paper, but it was quite amusing while it lasted). The great thing that it did was to stimulate a number of people to check much more carefully than had ever been done before what in fact the ratio of the two charges was. This was because we had shown in our paper how big the difference had to be if it were to have any significant cosmological consequences, and so the experimenters knew what precision they had to aim for to get a significant result. The stimulation worked beautifully. As a result of that paper a number of experimenters (and I think Hughes did it with the greatest precision), measured the ratio of the two charges. They found that it was much closer to minus one than anybody had known before (if it was not exactly minus one, but one cannot tell that experimentally of course). In fact, with what Lyttleton and I had worked out it was quite clear that any difference from minus one was too small to have any cosmological significance. I am not a bit ashamed of that paper. It is not that I had dragged science backwards. We stimulated something that led not to what we had expected—but there is more knowledge and more understanding as a result of it. As this little story exemplifies one is always relying on the interaction of theory and experiment. They are totally intertwined. Just as I said earlier that 'truth' was a word that had nothing to do with science, because any experiment is always so theory-laden both in the arrangement and in the way you look at it, it would be quite fallacious to regard any description of the outcome of any experiment as 'fact'. Indeed 'fact' is an emotionally loaded word which is entirely unhelpful.

To come back to the technological basis of science, it is a continual improvement that we are witnessing. I always find it intriguing to think that what is so often called the revolution in physics in the last two decades of the last century (what with the discovery of electrons and X-rays and so on) as far as I can make out owes its particular timing in history entirely to the fact that not only did reasonably reliable vacuum pumps become available then for the first time, but also (and I believe equally important) that plasticine became available to stop the leaks. The technological basis of scientific progress is thus very clear.

One can, of course, take a very different view—a cynical view if you like. Max Planck once said, 'It is not that old theories are disproved: it is just that their supporters die out.' An uncomfortable statement, but perhaps not a wholly incorrect one. After all, when I was speaking earlier about Newcombe's analysis of the orbit of Mercury disproving Newton's theory, I said that it was not exactly recognized at the time. The swinging away from Newton's theory came fifteen years later when the special theory of relativity was becoming fully accepted as a natural continuation of Maxwellian electrodynamics, and one just could not fit Newton's gravitation theory into the intellectual framework of special relativity. It was then that Einstein started to work on his theory of gravitation. It is quite interesting to see that in that framework one then looked first for a theory that should give much the same answers as Newton's theory, and secondly something that could deal with the discrepancy that had been found fifteen years earlier concerning the orbit of Mercury. Naturally, if one had wanted at all costs to maintain Newton's theory, there would have been other ways of dealing with that discrepancy. There might have been a planet yet closer to the sun—it was even given a name, 'Vulcan'—exerting such perturbation on Mercury's orbit as led to the observed consequences. There might have been a lack of spherical symmetry in the sum that would have led to similar discrepancies. One of the most interesting variants of Einstein's theory of gravitation, the Brans–Dicke theory, says that in fact not all the discrepancy between Newtonian theory and observation should be explained by a different theory of gravitation, but it should in part be accounted for by a lack of spherical symmetry of the sun. I am just throwing this in to show how we are always in a fluid state. Hard experimental tests combine with the intellectual discontent, intellectual movement, intellectual attitudes of the moment. Certain kinds of theory, certain kinds of explanation, are then considered acceptable, and others are not considered acceptable. One might say this is true also outside the field of natural science.

Three hundred years ago most people—very many people at least—believed in the existence of witches. Very few people now believe in the existence of witches. But to the best of my knowledge nobody has ever *disproved* the existence of witches. It is just that they do not fit into the kind of intellectual climate in which we live. The progress, the leap-frogging, curvey progress of knowledge, is one that has not just technological and explicit theoretical strands, it has a good deal of underlying mental attitudes formed by what has gone before and affecting what becomes acceptable. Sometimes attitudes that are possibly underlying particular points in theories and outlook are referred to among the cognoscenti as 'folklore'. There is a good deal of such folklore in physics, and in our interpretation and understanding of physics: a great deal of mythology even, as I said on a different occasion in another university city. A good example of this is Maxwell's theory which, as you know, grew out of Faraday's ideas, out of the ideas that electromagnetism must be expressed in the form of a field. The extraordinarily successful aftermath of Maxwell's theory led people to believe that the presupposition, the intellectual attitude of looking at fields, was obviously so superior to anything else, that everything had to be put into that and the world *was* fields. It was Wheeler and Feynman, twenty-five years or so ago, who showed it was in fact possible to put Maxwell's theory into an action-at-a-distance formulation. While the field point of view undoubtedly did help Maxwell to find the equations, and did help people to understand them, it would be quite wrong to say that the tests that Maxwell's equations have stood are the tests that the field concept has stood, because you can arrive at the equations otherwise. There is, then, always this underlying miasma which is very important to us in the way in which we communicate with each other, which is a very essential part of science. Communication is essential. If you just have great insight and keep it to yourself you might be a mystic—you are certainly not a scientist. In communication a common intellectual background is very important. Even with it there is great difficulty in understanding scientific papers. I am reminded of a remark Freeman Dyson is said to have made once, that it is the duty of an author of a paper to make the subject matter clear to at least two people, one of whom may be the author. This is not always achieved.

It is difficult to talk of the idea of progress here without referring briefly to an old chestnut in science, namely the question of the direction of time. It is well known that Newton's equations and Maxwell's equations are completely reversible and that on the macroscopic scale there is therefore no reason why things should go one way in time rather than the other way in time. (But for one obscure possibility, time symmetry applies to microscopic physics too.) It is not at all easy to see where the uni-directionality of time comes from. It is fairly straightforward to follow through it to three separate roots. There is the thermodynamic root of the asymmetry of time, which in the simplest terms can be exemplified in the example of a basinful of water that is hot at one end and cold at the other. It will all be lukewarm after some time, but it is never observed that lukewarm water will separate into hot and cold without your putting the refrigerator to work. That is one root of the direction of time. A second origin of the direction of time occurs in Maxwell's electromagnetic theory. In that theory an accelerated charge is compatible with a variety of radiation fields. This variety consists of a linear superposition of what we call the wholly retarded solution that it is a wave going out from the charge, and an advanced solution that it is a wave that converges on to the accelerated charge. In practice we observe only the retarded solution. This is a second root of the direction of time. A third root is that we observe a red shift of the distant galaxies and not a blue shift.

The question of the direction of time (and I can hardly talk about progress without mentioning the direction of time) is perhaps in the main the problem of whether these three are closely linked—are in fact all derivable from some one principle. A certain amount of grey matter has been poured into this without very convincing answers, largely because it is extremely hard to think of tests that you could actually perform. But I will mention one point, namely that if it were possible to bring these three together by basing ourselves on a feature of the universe—that in an expanding universe, a universe with red shifts—temperatures will tend to equalize and radiation will travel outwards from the source—if this were possible, then the question of the direction of time is solved. You would not then say—'Ah, yes, but the universe could be contracting', because in that case, if you had a contracting universe, all these other processes would be reversed too. If in such a contracting universe you had people like you or me, the way their minds would work would also be reversed, so that they would have a reasonably good memory for the future and an appallingly bad memory for the past. So they would naturally use the labels 'past' and 'future' in the opposite sense from the one in which we use them. Hence these people would perceive that they lived in an *expanding* universe. So it is not necessary —this I think is the point—first to say the three are connected, and then, that the universe is expanding. If the three are connected, the universe is expanding by definition. This is a step which does not get you all that far because the intermediate area is in a mess, but at least it suggests that a sudden reversal of the meaning of progress is unlikely because the time-constant of the universe is fairly large. Cosmology is a field like others in which progress would be measured by having theories that are scientific in having testable consequences. However, progress here consists often not so much of progress *in* science, but in bringing something that before was not science and making it part of science. This is of course, something we have seen happening in various areas, and I trust it will happen a good deal more.

If I may then sum up, it is quite possible in looking at the history of science to find a direction in it. If you wrote a backwards history of science, it would read very oddly indeed. In that sense there undoubtedly is a progress in science, but if you ask 'Is it true that today we feel *in all respects* more comfortable about the extent of our knowledge than we did yesterday?' then the answer is almost certainly 'No'. My own deduction from this is that whatever politics and sociology may bring us to, whatever may happen about support for science, about some areas of science becoming forbidden, or people being enthusiastic about them, one thing that I am certain of is that science will not become dull.

2 On the molecular theory of evolution

J. L. MONOD
University of Paris

IT is a great honour indeed to speak in this University on any subject, but I am particularly sensitive to the fact that this is the Spencer Lecture. It recalls some very old memories of mine. I was raised in the ideas of a father who was born around 1860. He belonged to a generation engrained with the ideas of what we now call the Scienticism of the nineteenth century. His great heroes were John Stuart Mill and Spencer. He, as they, believed strongly that there was something called Evolution, which was written into the structure of the universe; and that there had been, and would be, an automatic, necessary evolution of knowledge and science, which would go hand in hand with ever-better societies. Increase in scientific knowledge and attitudes would, as it were, secrete better societies. Of course, we know better by now. But still, I would not readily abandon these ideas completely, even though particularly if we look at some of Spencer's ideas on evolution, we must admit that they were extremely naïve.

What I wish to talk about today is the present state of the theory of evolution. Let me say right away that when I talk of the theory of evolution, I am speaking strictly of the theory of evolution of living beings within the general framework of the Darwinian theory, a theory which is still alive today. Indeed, it is even more alive than many non-biologists may think.

The theory of evolution is a very curious theory. To begin with it is necessary to recall that in many respects the theory of evolution is the most important scientific theory ever formulated, because of its general implications. There is no question that no other scientific theory has had such tremendous philosophical, ideological, and political implications as has the theory of evolution.

It is also a very curious theory in its status, which is quite different from that of physical theories. The basic aim of the physical theories is to discover universal laws, laws which apply to objects in the whole of the universe, with the hope of being able, from these laws—that is to say from first principles—to derive conclusions, explain phenomena throughout the universe. When a physicist looks at a particular phenomenon it is with the hope that he will be able to show that he can deduce this phenomenon from universal laws, from first principles. The theory of evolution, by contrast, has a different aim. It has a range of application, which is not the universe, but only a tiny corner of the universe, namely the universe of living things as we know them today upon the earth. We can define the aim of the theory as that of accounting for the existence today on the earth of about two million animal species and about a million plant species, plus an unknown number of species of bacteria.

This is a very small corner of the universe, and it is very dubious whether the existence of these very peculiar objects—living beings—can, or ever could, be derived from first principles. I might say now that I do not believe it will ever be possible to do such a thing, for very profound reasons that I will try to explain.

Another curious aspect of the theory of evolution is that everybody thinks he understands it. I mean philosophers, social scientists, and so on. While in fact very few people understand it, actually, as it stands, even as it stood when Darwin expressed it, and even less as we now may be able to understand it in biology. In fact, the first great misunderstander was Spencer himself. He of course was one of the first great evolutionary philosophers, but he was also the very first to show the inadequacy, as he said, of selective evolution to explain evolution.

The other great difficulty about the theory of evolution is that it is what one might call a second-order theory. Second-order, because it is a theory aimed at accounting for a phenomenon that has never been observed, and that never will be observed, namely evolution itself. In the laboratory we are able to set up conditions so that we may be able to isolate mutations of a given bacterial strain, for instance; but to observe a mutation is a very far cry from observing actual evolution. That has never been observed even in its simplest form—the one which is required by the modern theorists to account for evolution, namely the simple differentiation of one species from another. This is a phenomenon that has never been seen. I would not say it never will be, but it seems extremely doubtful. Therefore, if you look at the structure of the theory of evolution, if for instance you open one of these great modern books about evolution, such as the books by Dodzhansky or Meyer, or Simpson, you will see that the discussion always goes the following way—that one starts from the actual data, that is to say the present structure, performances, and anatomy of a given group of animals, and then one looks at the fossil record, and from the fossil record and classical considerations of comparative anatomy, one tries to derive filiation of these forms. With the help of modern biology, such filiations can be also reconstructed by the analysis of sequences in certain molecules such as proteins. What has happened is that the molecular filiations have very beautifully shown that the anatomists were right. Filiations reconstructed in this way are perfectly coherent with filiations reconstructed by the anatomists. So the first thing you do is propose a first-order theory, to answer, for instance, 'How do men descend from fishes?'. And you build up a certain set of filiations. And then comes the second-order theory. You want to account for these facts and you introduce all sorts of further considerations, consisting of assumptions about rates of mutations and so on that might have occurred. And finally a reconstruction of the ecology of the groups that are assumed to have come before is required, because in order to account for evolution in terms of the selective theory, you have to assume some sort of selective pressure. Selective pressure is something that develops according to the milieu in which an organism develops, but also according to the performances and, I would say, the personal preferences of the individual. Once you know all that, you have the conditions for a particular application of the theory of evolution to, say, the evolution of man from some fish somewhere in the early secondary era. This is what I mean by second-order theory.

Clearly, no reconstruction of this kind can ever be proved, and worse than that, no reconstruction of this kind can ever be disproved. This might appear to make the whole selective theory of evolution an extremely intellectual construction, especially in view of an epistemology such as that of Karl Popper for whom—and I completely agree with him—the distinctive mark of a truly scientific theory is not that it can be proved—because no theory ever can be proved right—but that a scientific theory must be of such a structure that it can be *dis*proved. In any particular instance this is not the case for the selective theory of evolution. However, this difficulty is not really peculiar to the theory of evolution. Many great physical theories which are corroborated and accepted for their general contents may be very difficult to apply in any given instance. This is the case, for instance, for quantum mechanics. Once one tries to apply quantum mechanics to the description and prediction of the properties of individual chemical molecules of any complexity—even relatively simple ones—one gets into extreme difficulties and is forced to adopt rather arbitrary types of reasoning. However, we accept the quantum theory as forming the basis of all our understanding of chemistry, for different reasons—that is to say, for its general content.

Now what has not always been fully seen is that in fact the selective theory of evolution, even as it was first proposed by Darwin in the first edition of *The origin of species*, has (in fact had, at the time), predictive contents much richer than Darwin himself knew or ever found out. Let me give you two examples. What I mean by 'richer content' is that once the theory of selective evolution has been formulated, as Darwin did in 1859, then a certain number of consequences must follow, even though the author of that theory—Darwin in that case—did not (in his days hardly *could* have) seen these consequences. The consequences sometimes go far beyond the selective theory itself or even biology itself. I think one of the most remarkable examples is the famous discussion between Lord Kelvin and Darwin—a discussion which Darwin thought he had lost. Darwin was very much aware of the fact that in order for his theory to be acceptable one needed an enormous expanse of time for the evolution of living beings to have taken place, and he spoke in terms of hundreds of millions of years—many hundreds of millions of years, without being able, of course, to give any sort of precise figure. Kelvin, who was both one of the greatest physicists of his time, the greatest thermodynamicist in any case, and also a deeply religious man (which may have had something to do with his attitudes), proved, by his calculations, that the life of the solar system could not possibly have exceeded say about twenty-five million years. This clearly was not enough for Darwin, and almost reduced him to going back to other interpretations of evolution itself. If there are any Marxists here they will be happy to learn that this can be explained on Marxist grounds, because Kelvin, though very religious, was a great scientist of nineteenth century England, and had as his model for the energy of the sun, a coal pile. He had no other choice, and by calculating the dissipation of energy from a coal pile of the size of the sun, he could conclude that it was impossible to assume that it could live for more than twenty-five million years—which is pretty good for a coal pile. As we know, even our fuel is not going to last that long. Now, of course, he was wrong, and we know now that solar energy derives from nuclear energy, from fusion in fact, so that we might say that the discovery of nuclear energy, or fusion, or more truly of the famous Einstein equation relating matter to energy, was implicitly contained in the selective theory of evolution of Darwin. This is curious, but it is a fact.

Let me give you another example, which is much more directly related to biology and to the history of the theory of evolution. Around 1871, a mathematician from Edinburgh proved mathematically that Darwin's selective theory of evolution could not possibly be right on the basis of the accepted ideas of the time concerning inheritance—heredity. These accepted ideas that nobody discussed—not even Darwin—were that inheritance was essentially a system of blending and that the offspring from a given couple was a sort of dilution of traits coming from each parent, and that this would go on in successive generations. By fairly simple calculations Jenkins showed that even if a new trait were able to appear by some sort of mutation, 'sport', as Darwin used to say, it would be diluted very fast in the populations which shared, or were destined to share, this heredity, and that therefore within the next two or three generations it would be diluted to the point where it would not possibly have any selective value. Jenkin's reasoning is absolutely without answer. Darwin, I think, probably realized that. It is the reason that in the last edition of his great book during his lifetime he went back on the selective theory, and accepted more and more of the kind of Lamarkism which he had tried to tone down in the first edition.

What Jenkin's remarks called for was a theory of heredity by which inheritance would be essentially discreet, discontinuous, and ensured by units that could be transmitted from generation to generation without losing their somatogenic qualities. Such is the gene. The gene was discovered by Mendel during the lifetime of Darwin, but of course it was not scientific knowledge at the time. It remained virtually unknown until the beginning of the twentieth century. In fact, strictly even more than in regard to Kelvin's objections, the selective theory of evolution as Darwin himself had stated it, required the discovery of Mendelian genetics, which of course, was made. This is an example, and a most important one, of what is meant by the content of a theory, the content of an idea, the content of a *logical* idea concerning the universe, concerning the world. If it fits, a good theory or a good idea will always be much wider and much richer than even the inventor of the idea may know at his time. The theory may be judged precisely on this type of development, when more and more falls into its lap, even though it was not predictable that so much would come of it. That is what I would like to show now in terms of the more modern aspects of biology.

As probably most of you know, as a result of the discoveries of the first half of this century in genetics, there developed a new theory of evolution which integrated the general ideas of Darwin about selection with the whole of what was known at the time of classical genetics. This theory has been expounded by many writers in this country especially; the most important people have been Fisher, Haldane, and Sir Julian Huxley. It is to Sir Julian Huxley that we owe the synthetic theory, as he liked to call this modern concept of evolution. I would like to give a very short résumé of the basic ideas of the synthetic theory, and then show that later developments in molecular biology in fact again completely fall into the lap of this new concept of evolution—but not entirely new, because it depends entirely on Darwin and is a development of the Darwinian theory.

This can be expressed in three statements.

1. According to the modern theory—the synthetic theory—the units of transmission, the units of conservation in heredity, and the unit of mutation are the same, namely the gene. I call it the unit of transmission and also the unit of mutation.

2. The second basic statement is that the unit of selection is, as Darwin thought, the individual—not the gene. It is the individual that is the unit of selection, but not through the rather naïve process that appears in the writing of Darwin himself, or Spencer or others—not through the rather ferocious and all too simple concept of struggle for life or survival of the fittest, but in the form of selective pressures which modulate the probability that genes coming from this individual will be found in the next generation after one, two, three, four, or n generations. What counts for selection in this concept is not the survival of the individual which is, by itself, of no interest to selection, but whether his gene will have a chance of being present in

the first, second, third, fourth, nth generation. That is what we need in order to have evolution, and to have selective pressures that mean something to evolution. The survival of an old man, if he is quite unable to have any children, is of no importance whatever to the theory of evolution. It will not contribute to the evolution of man, even if he is very old and very bright.

3. The third concept, which is in many respects the most important and the most misunderstood, has been the most elaborated upon. It is that the unit of evolution is not the gene, nor the individual, it is the population. But not any population. It is that population which shares a certain genome. That is to say, a Mendelian population with sexual recombination, leading to the steady production of new genomes by recombination. It is a unit which actually is able to evolve. As you may know, a great deal of theoretical thinking has gone into this concept, including the developments of complicated and very interesting models of evolving populations.

This is a very brief résumé of the synthetic theory. Let me again say that this is, you might say, the first straight outgrowth of Darwin's theory once it had integrated and assimilated all the results of classical genetics.

The last adventure in biology is the sum of discoveries and knowledge which is summed up under the term 'molecular biology'. I would like to try and define something which I will call 'the general molecular theory of the genetic code', because I wish to use the phrase 'the genetic code' not in the restrictive sense in which it is often taken, namely the mechanism and the symbolism by which a sequence on DNA is translated into a sequence in protein (a sequence of nucleotides is translated into a sequence of aminoacids), but the whole complex and precise knowledge which we now have as to how the genome exercises its two basic functions—namely, first to replicate itself true to type, and secondly to transfer its information and its 'orders', as you might put it, to the cell and the individual.

We could summarize the genetic theory of the code in the following way. There is the gene, which as you all know, I am sure, is a double strand of DNA (or RNA in some cases). Now this is able to replicate, to make more genes, which are copies of the one from which they stem. Then, genomes in a population which shares these genes may undergo mutations of individual genes, which modify in a discrete way the information which was contained in the original genome. They undergo recombinations so that novelties which appear as mutations may be tested one against the other—that is to say novelties which have appeared in different individuals in the population will now be reassembled, de-assembled, and tested in the form of new genes. It is easy to see that as soon as you have as few as say a hundred different avenues for the propagation of genomes, supposing that there are two different forms with slightly different information for, say, a hundred genes, the number of combinations that can be reached is enormous. This is the source of all novelty, according to the theory.

Now the genome also has to do things and express itself in the cell. This we know it does by a complex mechanism. There is first an event called 'transcription' by which the genome is transcribed into a sequence in nucleic acid. Then this sequence in nucleic acid is converted into a corresponding sequence in a protein molecule. This process is called 'translation'. Thus there is a sequence of a certain chemical species—nucleotides, then there is a sequence in another type of chemical, namely amino acids.

The sequence of amino acids as first formed is still a biological object which has no recognizable function, and which is completely inactive. In order to perform a function it has to fold up into what we call protein. This is a complicated folded-up structure—the kind of thing that Professor Phillips studies so brilliantly. Now only at this stage does what we call the meaning of a given gene appear in a complete way that we can study, measure, and so on. The proteins can be considered to have several, in fact all, the basic functional activities that we need in order to understand the functioning and the development of living beings. Some have catalytic functions, some have regulatory functions, some have both, and some have morphogenetic functions.

We also know that these regulatory functions are of tremendous importance and that it is due to the regulatory functions of proteins that an individual is a whole coherent system working within itself, for itself we might say, due to the enormous number of feedback loops between proteins and other functions in the cell, in particular, feedback loops which go back to the gene. In other words, the genetic information by which they are translated into these active components *and* the result of the activity of these active components now inform the genome as to what to do or not to do. This is what I would call the essence of the molecular theory of evolution—all of it may be considered to be a further explication, development, and verification of implications which are expressed, or unexpressed but present, in the synthetic theory of evolution.

To begin with, and this is perhaps one of the most important points, the theory of evolution—in all its forms but especially in its more modern forms—assumes, explicitly or not, that there is a profound basic uniformity among living beings, that the basic machinery is the same in all. This is undoubtedly one of the great results of modern biology—I will not give you examples of it—there are too many. The best example is the fact that the genetic code itself, that is to say the chemical machinery of inheritance, works according to the same basic principles and the same code in every known living being from bacteria to man.

A second assumption of the synthetic theory is that all inheritance and therefore all morphogenetics—all that distinguishes one species from another, or one individual from another—must be referred to information in the genome. Since it is the only information contained in an individual that is transmissible to its descendents, it is the only information upon which selection can have any effects. That is a basic assumption which, you may be surprised to learn, easily could have been considered unproved, and was still disbelieved something like twenty years ago by many biologists. It has been completely accepted only as the result of the discovery of this transcription–translation machinery, and the results which showed that not only does the genome contain information concerning the structure of molecules, but is also a regulatory system, which is both a transmitter and receiver of functional information. This also was implicit in the synthetic theory, to the extent that the synthetic theory considered the population and therefore the whole gene pool as the unit of evolution, and because of the concept of an integrated genome— an integrated system whose properties had to be considered as a whole rather than unit by unit.

The two other conclusions which may now be considered proven are also in this scheme. They refer to the nature of mutations. Darwin had at first talked of sports which happened more or less at random. The geneticists of the classical period observed mutations, and included in the synthetic theory the idea that mutations were spontaneous events that were not controlled from outside. We now know what the nature of mutations is. We can even write chemical formulas for most mutations. We know that they are quantum events, that they occur at the level of single molecules, and therefore that they belong in the realm of microscopic physics—in the realm of events that by their very nature cannot be individually predicted and cannot be individually controlled.

This scheme tells us that the old idea of acquired characters, which had been proposed by Lamark, not only has never been verified, as you probably all know, but in fact is completely incompatible with all of what we know of the whole structure of transfer of information. First, the spontaneous nature of mutation is incompatible with such an idea and second, we know that this sequence of transfer of information is essentially irreversible. Here, I think, I must correct a wrong idea that has been spreading for the past three or four years. It was discovered some years ago that in some cases, the transcription step from DNA to RNA works in the reverse direction. That is nothing surprising. This is a very simple step and even by the basic principle in physical chemistry of the reversibility of microscopic events, it could be predicted that such events could occur. They do occur, indeed, but this must not be taken to mean that *information* from protein could possibly go back to the genome. I think, in spite of some hesitation even by some very distinguished colleagues, I am ready to take any bet you like that this is never going to turn out to be the case.

So you see that the advance of molecular genetics, of molecular biology, has in fact enriched tremendously, and made explicit a great deal that was implicit in the theory of evolution. It has not revolutionized this concept. On the contrary, it has, if anything, both made it much more precise and much harder. It is a harder theory in its description, it is a harder theory in the sense that it tells us a great deal more, and therefore becomes far more sensitive to the criterion of falsification of Popper. And I would say that we have a complete theory of evolution only now that we have a physical interpretation of these basic steps of phenomena that must be assumed, and that are known to account both for the development of living beings, for the inheritance of their traits, and for evolution.

The upshot of all this is that it is a conceptual error to say that evolution is a law, or even that evolution is a law of living beings. This is wrong. The privilege of living beings is not to evolve but on the contrary to conserve. (I'm sorry to say that, especially in front of a great crowd of undergraduates, but this is the case.) The privilege of living beings is the possession of a structure and of a mechanism which ensures two things: (*i*) reproduction true to type of the structure itself, and (*ii*) reproduction equally true to type, of any accident that occurs in the structure. Once you have that, you have evolution, because you have conservation of accidents. Accidents can then be recombined and offered to natural selection, to find out if they are of any meaning or not. Evolution is not a law; evolution is a phenomenon that occurs when you have structures of this kind.

Finally, I would like to discuss one of the more philosophical or ideological aspects of the theory of evolution. My main aim will be to try to understand why the *selective* theory of evolution strictly speaking, is still so widely attacked, doubted, misunderstood, and misinterpreted. One can classify the various scientific theories or ideologies or philosophies which either directly or by implication are incompatible with this theory as it stands. And I like to make a classification of these theories in three groups. First, the most classical dissenters are the vitalists. And if one is going to drop names, the first well-known one in our era is Driesch, followed by the French philosopher Bergson, and among contemporaries people like Wigner, Polanyi, and many others. The interesting thing about these contemporary dissenters is that they are all physicists, not biologists. Well, you all know what a vitalist is. He is somebody who thinks that physical laws by themselves are unable to account for the existence of living things and their evolution. So something more has to be found, something more has to be injected into the theory—in fact the physical laws either are not applicable or they are insufficient, and we have to discover something more. They are not really very interesting: let's leave them.

The second class are the animists, which in many respects are much more interesting, more profound, and relate to very old concepts that man has had of his own nature and his relationship with the universe. What the animist says is that evolution indeed is a law, that it is a universal law, not only of living beings, but that there is some sort of ascendant law of progress—differentiation, perfection, and so on—in the whole of the universe, from electrons up to animals and living beings. One of the first great exponents of this idea, I'm sorry to say, was Spencer. Besides Spencer, and about the same period, one can find a group of very important people indeed, Hegel, Marx, and Engels. There is no doubt that there is a profound affinity between the attitude of Spencer and that of Marx and Engels and Hegel—even though the theories that they put out are fundamentally completely different, it is perfectly clear that the basic aim of the theories, that is of the attitudes of these theorists, is that evolution is *the* law, evolution has a meaning, and evolution not only has a logical meaning, it has a moral meaning. That is really the basis of all these attitudes including, in particular, Spencer and the Marx–Engels system. I am thinking both of historical materialism, but more specifically of dialectic materialism.

Finally, in more modern times, there is a new, subtle class of animists—which I will call the thermodynamicists. They are not thermodynamicists like Lord Kelvin, even though some of them are extremely good indeed. They try to devise theories, formulations, formalizations, from which they think they can show first, that life could not have failed to start on the earth, and secondly that evolution could not have failed to occur. This is not a completely fair description. I am interpreting their motivation rather than giving you a résumé of what they actually say. They never dare say as much, but that is what they would like to say. In this category I will put Manfred Eigen. The first time I heard Manfred Eigen lecture on the subject of evolution, he was developing big formulations that I did not understand, which tended to prove that evolution could not have failed to occur. I was sitting next to a colleague of his and he turned to me and said, 'You know, the truth of the matter is that Manfred cannot conceive that evolution could have happened without the aim of creating Man . . . fred.'

Now I do not want to be too sceptical or severe but I honestly think that so far the attempt to interpret evolution in thermodynamic terms has led absolutely nowhere, has not enriched the theory of evolution in any way, and that all these attempts in fact fall foul of the principle of falsification. They are theories from which you cannot possibly derive any sort of observations or experiment that would prove them wrong. Therefore, they are void of any content and we do not need to discuss them any more.

What is very interesting, and it is with this that I should like to end this lecture, is why is there this constant resistance, this rejection of the theory of evolution, of the selective theory of evolution as we understand it. I think the interpretation is perfectly simple. It is a very old and profoundly engrained concept in man that anything that exists, in particular himself, has a very good, an obligatory reason to be there. The aspect of evolutionary theory that is unacceptable to many enlightened people, either scientists or philosophers, or ideologists of one kind or another, is the completely contingent aspect which the existence of man, societies, and so on, must take if we accept this theory. If we accept this theory, we must conclude that the emergence of life on the earth was probably unpredictable before it happened. We must conclude that the existence of any particular species is a singular event, an event that occurred only once in the whole of the universe and therefore one that is also basically and completely unpredictable, including that one species which we are, namely man. We must consider our species as any other species— we are a single species, a single event—and therefore we were un-predictable before we appeared. We are completely contingent in respect, not only to the rest of the universe, but even in respect to the rest of living beings. We might just as well not have been there and not have appeared.

This is even more true of man than it is of most other species, because as it turns out man is endowed with a completely unique capacity, which no other species shares, namely language. I think all linguists and ethologists now agree that there is a huge gap be-tween animal communication of any sort and actual human language, by which we are able to express essentially logical, symbolic, and, as Popper puts it, argumentative thinking. There is nothing argumenta-tive for instance, in animal communication. The mere fact that not only are we a single species, in spite of the existence of different races, not only do we have for all we know a single origin, but we also have this unique competence that has never appeared in any other corner of the known biosphere. It is therefore again a completely unique event and again one that could not possibly have been deduced from any sort of first principles and certainly not from the general theory of the structure of living beings and of evolution.

Finally, this unique event led to a completely new outlook in the universe, something which can be compared in its novelty only to the appearance of living beings from an inanimate system, namely the appearance of what Karl Popper calls—World Three, and by this term he means the abstract world of ideas, of logic, of mathe-matical objects, of arts, or creativity. He argues that the existence of the third world as an entity in itself needs to be recognized, and he pleads against the psychologies of much of modern philosophy, to show that this abstract third world, which in French I like to call '*royaume des ideés*'—the kingdom of ideas—merits the word 'exist-ence' predicated of it, even though it is not an easy thing to define. We do not need to define it. But if we want single examples, we cannot deny that, say, the fifteenth quartet of Beethoven exists, even when it is not heard, as a logical construction, as a structure, which can eventually be transmitted in different ways, through our brains or through computers or through tapes and so on. We cannot deny that mathematical entities like, say, irrational numbers, exist in some ways. They are the products of the existence of man—of his having acquired the symbolic capacity to communicate and of the development of this capacity. Still, the world of numbers exists with a large degree of independence. We have examples of this in the history of the theory of evolution which I have just outlined, in the fact that certain ideas had an obligatory content which was much wider, and was yet there, even though the inventors of the ideas did not know it and it was not present in any single man's ideas. I mean to say, for instance, when Darwin published *The origin of species*, nobody had any idea of discreteness and invariance in heredity, and yet we see now that this idea was implicitly contained in the theory of evolution of Darwin. Similarly, when our ancestors, pure savages who could not write, but could communicate and could count—at least on the fingers of their hands—built up a system of numeration which, although of course they had no idea of it, contained both odd and even numbers, irrational numbers, and so on.

This, of course, leads to a tremendous question which I will not try to answer. 'How is it that the abstract third world of logic, ideas, scientific theories, and mathematics—how is it that this third world turns out to be adequate to the description of the real world?' This is really, I think, the most fundamental philosophical question.

3 *Biomedical advances: a mixed blessing?*

W. F. BODMER
University of Oxford

I FEEL greatly honoured to have been asked to speak in a series of Herbert Spencer Lectures and so become part of such a long and distinguished list of lecturers. The challenge presented to me I find particularly daunting as I appear to be the only representative of the home team. My task is lightened somewhat by the two previous contributors who have so elegantly laid the groundwork of this series, and particularly by Professor Monod who has introduced the basic subject of genetics from which I shall, as I am a geneticist, be drawing most of my examples.

I subscribe to the general idea that knowledge, and in particular scientific knowledge, generally increases and so, in some sense, science inevitably progresses and has an over-all direction. This progression is analogous to evolutionary change which is also generally irreversible. However, just as in the case of evolution one cannot say that all evolutionary change (which is mediated by the process of natural selection and depends on differences in reproductive fitness) is necessarily, in some Utopian sense, for the good of our society as Herbert Spencer supposed to be the case, so also must one deny that advancing knowledge is necessarily for the good of mankind. Applications of scientific advances in society nearly always create tensions and conflicts. It would indeed be a comfortable world, especially for the scientist, if Herbert Spencer were right and we could simply let advancing evolution and science, going hand in hand, inevitably produce progress for the greater good of mankind, without our being concerned as to how they managed to do so. Isaiah Berlin in his essay on historical inevitability has pointed out how determinism in human affairs, especially if interpreted literally, severely limits, if not destroys, the concept of individual responsibility, and he has emphasized how dangerous this could be. There is an analogous danger in ignoring the tensions created by the application of new scientific knowledge and assuming that scientific progress is necessarily for the good of the society, however that good be defined. As Isaiah Berlin warns us there is danger in assuming 'that men will know more and therefore be wiser and better and happier'.

I believe there is an important dichotomy between basic scientific advances and their application. To my mind this is exemplified by the differences in attitude between basic scientists and, for example, politicians, doctors, or lawyers, when they are presented with a problem. The scientist will only choose to attack the problem if he believes it to be soluble using available methodology. In the words of a famous previous Herbert Spencer lecturer, Peter Medawar, science is 'the art of the soluble'. The lawyer or doctor, on the other hand, when presented with a problem; whether or not to prosecute, how to sentence, how to treat a patient, whether to operate, where to cut, has to make a decision, however inadequate the bases for doing so might be. This dichotomy is a little like that between C. P. Snow's two cultures, only in my view it is much more fundamental. It was brought home to me in a discussion with a group of lawyers on the rights of the unborn child and the time to use as the legal start to life. For a biologist this is clearly fertilization, or even before—all cells are in some sense alive—but for a lawyer this answer is impracticable. The lawyer has to make a decision which will work in society and thus, though as R. A. Fisher once said, 'birth may be a rather arbitrary start to one's life', it is the obvious time-point for the lawyer to use.

My main concern in this lecture is to discuss the contrast between basic science and its applications, to illustrate the tensions that are created by its applications with some examples from my own field of interest, and to emphasize that the tensions must be faced and cannot be avoided. The flow of knowledge will not be stopped or reversed, and even if it were, no problems would thereby be solved.

Genetics is an area that in many people's minds raises rather frightening social problems for our society, fears which are fed by such journalistic phrases as 'genetic engineering' and 'test-tube babies'. This was brought home to me rather vividly when I was returning with my family from a holiday on the Continent the summer before last. As we were confidently driving out along the 'Nothing to Declare' customs lane we were stopped by a rather severe-looking Customs Officer. Having seen from my passport that I was a Professor, he asked me first what University I was at— 'Oxford', I said—confidently hoping that that at least would placate him. However, next came the question—'What's your subject?'— 'Genetics', I said—silence for a brief moment and then—'Hm, I'd better let you go then hadn't I, otherwise you might turn me into a frog!'

I find this story elicits two very different sorts of reactions. The first is surprise that a Customs Officer should know about genetics and what it was, let alone especially in Oxford, its connection with frogs and, presumably, nuclear transplantation. The existence of such knowledge among 'lay' people must surely be considered as progress, at the very least in the important matter of disseminating science to society. The second reaction, which I must admit was my own immediate reaction, is one of concern that there should be such fear of the geneticist; almost as if he or she were the modern sorcerer, or at best, witch-doctor. This attitude surely represents a basic obstacle to progress in science, certainly of its application and possibly even of its execution if there is a consequent restriction on the freedom to do any research.

Such concerns about biomedical advances (and surely about other advances as well) are of course nothing new. Tissue grafting was recorded by the Egyptians as early as 3500 B.C., though the first proper description of some of the basic techniques was apparently due to Gasparo Tagliacozzi, a professor of anatomy at the University of Bologna in the sixteenth century. The theologians of his time bitterly attacked him, accusing him of impiously interfering with the handiwork of God and attributing his success to the intervention of the 'Evil One'. Towards the end of the eighteenth century the University of Paris banned all such types of operation. A few years ago here in Oxford, Professor Henry Harris, with John Watkins, developed a procedure for efficiently fusing cells from widely differing species, such as man and mouse. From these fused cells one can grow hybrid cells which are now much used in genetical and developmental studies, and in cell biology research in general. One response to Harris and Watkins paper was a cartoon showing Disney-like characters, including Mickey Mouse, sitting in the London Underground reading the *Daily Mirror*, with a caption

'Who was Walt Disney, Dad?'. Henry Harris has contrasted this cartoon with one published 163 years earlier in response to Jenner's research on vaccination using cow pox. Jenner is shown vaccinating a group of people, and those already innoculated have grown parts of cows on their arms and faces. The caption reads 'Cow pox. The wonderful effects of the new innoculation.'

The fact that such concerns are not new does not mean that they can be ignored. On the contrary, the increasing rate of advance in knowledge means that accompanying concerns arise more and more frequently and become more and more pressing.

The examples of biomedical advances I shall discuss will not include topical areas like genetic engineering, test tube babies and *in vitro* fertilization, and cloning. The days when, as the satirical columnist of *The San Francisco Chronicle*, Arthur Hoppe, said in response to a somewhat futuristic discussion by Joshua Lederberg, 'genetic engineers will be able to give the unborn child whatever characteristics mum and dad are praying for' are still, fortunately perhaps, a long way off. I should like to concentrate on issues that to my mind raise as many problems, but which are with us today. Better perhaps to deal with today's problems first, before solving those of tomorrow.

Applications of biomedical (and presumably other) advances fall, I believe, into two categories. The first can be called direct and include, for example, the discovery of vaccines, of penicillin, the importance of hygiene, and the development of surgical operations, all of which have direct effects on our health and the physical improvement of our lives. The second type of application is indirect and relates to discoveries which provide insights into the nature of human society that may help in the understanding of its problems.

The first example I should like to discuss, of direct applications, arises from advances in our understanding of genetic diseases and their detection, especially in the foetus *in utero* during comparatively early stages of pregnancy. A large number of diseases are now known that are simply inherited, most of which are extremely rare, occurring with a frequency of less than 1 in 50 000, and a certain proportion of which are well defined biochemically and can be traced to defects in single enzymes. Their biochemical effects can usually be detected in the blood or in cells grown in culture. These diseases are called 'inborn errors of metabolism' and were first so defined and studied by Archibald Garrod who was at one time Regius Professor of Medicine here in Oxford. For most of them there is, so far, no cure. But the question must be raised as to what can be done about them given the knowledge we now have.

One very well known genetic disease, which is quite common in some populations, is sickle-cell anaemia. This disease, caused by an abnormality of haemoglobin, occurs with a frequency of up to 2–3 per cent in parts of West Africa. Like all so-called 'recessive diseases', affected individuals carry two abnormal genes. However, most of the abnormal genes in the population (which when present in duplicate cause the disease) are found in 'carriers' who have one normal and one abnormal gene, and who are clinically normal and quite healthy. In fact, in those parts of the world where sickle-cell anaemia is common, the carriers have an increased resistance to malaria which most probably accounts for the high frequency of the disease in those areas. Individuals with the severe anaemia, who can only come from matings between carriers, mostly die in their teens, though improved over-all medical care is keeping them alive for longer and longer. There is, however, no specific cure for the disease. What solutions can one suggest for dealing with this extremely important health problem? Since all affected individuals are offspring of matings between two carriers, one possible solution is to try and prevent matings between carriers from taking place. To do this it would be necessary first to screen individuals to see whether they are carriers. Then, perhaps, one might hope that once they know they are, this knowledge will prevent them from mating with other carriers. A more radical approach might be to stop all carriers from having children, which would, 'at a stroke' so to speak, remove the gene from the population. In fact, preventing carriers from mating with each other would not be too severe a restriction on their liberty. More than 75 per cent of the population would still be freely available to choose from. The real problems with these approaches arise in putting them into practice in a population which may not fully comprehend the issues involved. The fact that the disease is one that occurs in the black population naturally creates the fear of a racial stigma, especially in the United States where it has been much discussed. Screening has to be done on the blacks and they may not understand why this should be so. Sickle-cell disease in the United States has in fact become a major 'political disease' and the subject of much litigation. In cases where screening has been carried out, carriers may find it hard to get medical insurance and feel stigmatized. Laws have been passed which do not distinguish between carriers of the gene and the diseased individual, only to add to the confusion. It is perhaps a little sad to reflect on the fact that this disease, the first one really to be understood at molecular level, has not, so far, itself benefited greatly from application of this knowledge, though it has contributed enormously to our understanding of many basic biological problems. Tensions created by attempts to apply the new knowledge have, I believe, arisen mainly as a result of misunderstanding and of inadequate education of the general public to appreciate the role of screening programmes.

Another well-known recessive disease in European populations is phenylketonuria, or PKU, which is a form of mental retardation caused by accumulation of toxic products from birth onwards. The mental retardation can be more or less prevented by appropriate diets, provided the disease is detected early enough. Its frequency in the population is between 1 in 10 000 and 1 in 20 000. There now exist very simple procedures for screening blood to determine whether an individual has PKU and, in fact, the majority of births in the United Kingdom are now so screened. Clearly, such screening programmes are quite expensive and the question must be raised as to whether they are worth having. At least a partial answer can be obtained from a cost–benefit analysis which compares the cost of screening and treatment, estimated to be roughly £4000 per affected individual, with the cost of maintaining untreated cases in an institution for their lifetime, which is estimated to be £24 000 or more. Thus, in this case even in these crude monetary terms, the answer to the question is obvious: the screening programme clearly is worthwhile. But, nevertheless some important questions do arise. The first is whether the cure is in fact complete. Are the individuals maintained on a diet completely normal? Do they have no IQ deficit whatsoever? The second question concerns the problems that may arise if treated PKU females have children, since, it is known that if they are not on a diet during pregnancy, all their children will be 'phenotypic' phenylketonurics. Thus, treated individuals must be followed during their lifetime to prevent this problem from arising. A third problem is that not all cases detected by the screening procedures are really PKU, and there is some question as to whether those that are not, and they may be hard to recognize, may not suffer from being put on the special diet.

Clearly the cost–benefit ratio advantage for a screening programme decreases as the frequency of the disease decreases. Most recessive diseases are so rare that it is hard to believe that a screening pro- gramme will be worth the effort in simple monetary terms. One must also take into the account the fact that, if there is even the slightest risk of detrimental effects of the screening procedure, when applied to a very large population this could lead to a significant number of cases. One must surely keep any detrimental effects created by the screening procedure down to a frequency that is less than that of the disease one is attempting to cure. With all these considerations in mind, how is the cost of human suffering and the advantage of its alleviation to be built into the cost–benefit equation? In a society with unlimited resources, one would clearly treat all such diseases, however rare they might be. In a society with limited, finite resources, one is compelled to make a choice on some criterion, and this inevitably leads to a cost–benefit analysis, whatever its deficits, and also to some consideration of the human component to be put into the equation in financial terms, however difficult that might be.

There is another inborn error of metabolism called Tay Sach's disease, that has been widely studied and which, like PKU, is well defined at the biochemical level. Infants with this disease weaken within the first six months, suffer from progressive mental and motor deterioration and eventually blindness and paralysis, and generally die within three years. In this disease, carriers of the abnormal gene are detectable. The disease occurs with a relatively high frequency, 1 in 2000, in Askenazi Jewish populations, while in virtually all other populations it has a hundredfold lower frequency. It is, because of its well-understood biochemical nature, one of the class of diseases that can be detected *in utero* following the procedure called 'amnio- centesis'. A sample of the amniotic fluid which surrounds the foetus can be taken at 14–16 weeks of gestation. This fluid contains cells which can be grown and used for biochemical tests or, if desired, can be examined for their chromosome constitution. When either of these tests reveals an abnormality, then one can offer the mother the possibility of abortion. In the case of Tay Sach's disease, cells from the amniotic fluid can be used to detect whether the foetus suffers from the disease. Thus, one way of guaranteeing normal children, as far as this disease is conerned, to matings between carriers who would otherwise have one quarter of their children affected, is to offer them the possibility of abortion when an affected child is detected. A screening programme among the Askenazi Jewish population of the Baltimore–Washington DC area has been instituted along these lines. Thus, wives are first screened to see whether they are carriers. If they are, then their husbands are also screened. Then, if both husband and wife are carriers, amniocentesis (the process of collecting and testing the amniotic fluid and its cells) is carried out on all their offspring, offering the possibility of abortion for abnormal offspring. In this way, clearly one can in principle avoid having any affected individuals in the population. The procedure is certainly cost-effective in the Askenazi Jewish population because of the high frequency of the disease, though it probably would not be so in other populations. This seems to be a rational and humane programme for dealing with the disease and so far it appears to be working well on a trial basis. Its success clearly depends on having a population which can appreciate the problems involved in such a programme.

There are at least two important questions raised by such programmes of *in utero* detection of genetic disease and subsequent selective abortion. The first of these is, of course, people's attitudes towards abortion. It is often objected to on religious and ethical bases. The second problem is the effect that such treatments might have on the eventual frequency of the disease in the population. In the case of recessive diseases, such as Tay Sachs or PKU, one can actually show that such effects will be very, very slight, since most of the abnormal genes are present in carriers who never produce affected offspring because they are not married to carriers.

Suppose now, however, that it becomes possible to cure a disease such as Tay Sachs. This could be achieved, perhaps, by enzyme therapy, namely by replacing the deficient enzyme much in the way that one gives insulin to a diabetic. Such a cure would undoubtedly be very costly and probably would have to be maintained throughout life. What now should be the decision? Should one cure all affected individuals, or should one continue a programme of selective screening and abortion? This latter is certainly likely to be the cheaper solution. What problems will be raised if the cure turns out not to be complete? It could quite easily result in a comparatively low IQ, that might well be difficult to detect for some time. Is it better not to be born, or to be born maimed or abnormal?

The question is often raised: would it not be possible to remove all abnormal genes from the population simply by preventing carriers of such genes from reproducing. But the point must be made that it can be shown that nearly all of us carry on average, up to four or more such abnormal genes in a single dose, where for the most part they go undetected and have no effect. If we prevent individuals who are carriers of those genes we can now recognize, from mating, we are indeed taking a very arbitrary step dictated simply by which particular diseases we now understand. If we took the more rational approach of stopping all people carrying abnormal genes from reproducing, this would leave few of us, if any, to reproduce. Clearly, any hope of eradicating all such genes from the population is quite unrealistic.

I should like now to turn to my second example of a direct application of scientific advances, namely, artificial insemination by donor, or 'AID', as it is often called. The first reports of artificial insemination appear to be of Arabs using the technique for horse-breeding in the fourteenth century, but for the most part the possibilities of its application in human populations arise from studies with economically important farm animals, particularly cattle. It is now possible to store sperm in liquid nitrogen at very low temperatures so that they remain alive indefinitely. AID can be considered as a treatment for infertile couples when the husband cannot produce fertile sperm. It is an alternative to adoption, and with the present drastic decrease in the birth rate which makes it hard to find children for adoption anyway, it is an alternative that is likely to be increasingly sought after. There are reports that children of AID are happier and more accepted in the home. Nevertheless, AID is subject to major legal and ethical problems. There have been two Royal Commissions in the United Kingdom, one reporting in 1948 and one in 1960 (the Feversham report), both of which were relatively hostile to the procedure. A more recent report, under the auspices of the BMA, (the Peel Report) which appeared in April 1973, was more favourable. Why are there these problems and what is their nature? Many centre on the question of the legitimacy of the AID child. Then there is the question of whether AID, if the husband was not informed, is a cause for divorce because it is tantamount to adultery. Should the sperm donor be identified, and what are his responsibilities if the husband dies prematurely? Does the child have a right to know who his or her biological father is, and should they in fact know whether or not they are a child of AID? At the present time, in fact, the AID child is not, and cannot be, legitimate, though most husbands presumably enter such a child on the birth certificate as if it were their own, thus actually perjuring themselves. This is a procedure that, in any case, must happen much more often as a result of 'ordinary' illegitimacy. Peel's panel, rightly in my view, recommended that the AID child should be legitimized and also that AID be available on the National Health Service. It is now, in any case, available in selected areas, depending largely on the decision of individual doctors.

There are two major questions with AID. Who should be donors? Should they be paid? Medical students have often been said to be a convenient source of sperm for AID, being good upstanding types with a minimal guaranteed IQ!

The BMA report naïvely suggests that sperm donors should be screened for genetic defects. But how is this to be done? As I have already pointed out, nearly all of us carry some abnormal genes in single dose. Are we all to be screened out? What characteristics, in any case, are to be screened for? Those which most of us are interested in, intelligence, behaviour, moral fibre, etc. are very complex and not simply inherited. We have no genetic tests for such characters, and the range of possibilities is so wide that no one in any case can agree on what would be the optimum. The geneticist Herman Muller, who was an ardent advocate of storage of sperm from eminent men for use in AID for eugenic aims, or eutelegenesis as he called it, suggested an answer to this by giving a list of famous thinkers, mainly scholars and scientists, whose sperm he would have liked to have seen stored. His first such list included Marx and Lenin. However, after a sojourn in Russia during which he witnessed Lysenko's ascendancy, Marx and Lenin were removed from his list, so even Muller did not find it so easy to choose.

Should one pay sperm donors? In the United States, in many cases, blood donors are paid but, as you know, in this country they are not. There is a strong tradition against payment, and we have a very fine blood transfusion service. What sort of people would one get to give sperm if they did so just for the money? Even commercialization of sperm banks seems to me a very dubious process. Such commercial sperm banks exist in the United States and the vasectomized male can for $18 a year have his sperm stored as an insurance policy against a desire to have children in the future. Such firms have plans to come over to this country and so far as I know there is nothing to stop them. Do we really want this? Is there anything bizarre in fertilization with sperm stored from a man now dead?

There is a report of a prominent American who has laid away his sperm to continue the family line in case his only son proves to be sterile.

Turning now to an example of indirect applications of science I should like to consider the question of individual genetic differences. Everybody looks different, apart perhaps from identical twins, but how much of these differences are genetic? The outward features, the characteristics by which we usually distinguish people, are not in general simply inherited. But there do exist a number of simply inherited traits, such as the blood groups (like the ABO system) and enzyme differences, for which common variants are found. These simple, genetically determined variations seem to parallel the differences we recognize between people. One of the most striking results of population genetic studies over the last ten years is the uncovering of just how much genetic variability exists in natural populations. Even using presently known genetic systems, the chance is less than one in a million that one should find two identical people. It is now even possible to give a very high probability for positive paternity, rather than simply paternity exclusion. Thus, one can now narrow down very much on the chance of a given child not having come from particular parents.

Transplantation of tissues illustrates, in a striking way, our genetic uniqueness. As most of you probably know, it is not possible, in the absence of appropriate drugs and matching procedures, to graft an organ from one arbitrarily chosen person to another and have it survive for anything but the shortest period of time. The reason that such grafts are rejected by the recipient is that there exist genetic differences between individuals which lead to the recognition of a graft as foreign by his immune system. These differences are rather like blood groups, and so the fact that, in the absence of appropriate medical treatment, all such grafts would be rejected is itself an indication that with respect to just these systems we are all genetically unique. In fact, matching for the more important systems coupled with appropriate drug treatment and careful medical therapy has now made transplantation of the kidney quite a successful medical procedure.

If the variation in the genes we now know and can detect is representative of all genes, then the potential for genetic variability is truly staggering. One can calculate that the probable minimum number of different types of egg or sperm that an individual can produce is $2^{10\,000}$, or $\approx 10^{3000}$, namely 1 followed by 3000 zeros. This can be compared with the total number of sperm that have ever been produced by all human males that have ever lived, which I estimate to be approximately 10^{23} to 10^{24}. Thus, the number of different types of egg or sperm that one individual can produce is some 10^{1000}fold more than the total number of sperm ever produced. Only a very, very small proportion of potential genetic combinations are ever realized. We are all genetically unique, and this genetic uniqueness clearly must apply to behavioural and other attributes as well as to blood type, though specific genetic factors that may be involved in these other traits have not yet been defined.

What are the implications of this variability for our society? Perhaps the most fundamental is with respect to the concept of equality. If equality really means the same, then it is clear that we are not all equal. One must then either say that equality means equality of opportunity, or else accept equal treatment for unequal people. One striking example of the conflict between genetics, with its demonstration of inherited differences between people, and political ideas lies in the former ascendancy of Lysenko in the USSR. Lysenko supported the old and outdated idea of inheritance of acquired characters, largely, perhaps because he felt that it was more in keeping with the political ideas of his country, since it implied that all people could readily be made equal by appropriate environmental manipulation (or perhaps because he saw this belief as a way to gain political power). He dominated biology in Russia and so suppressed its development along modern lines for some twenty or thirty years from the late 1930s, years which covered the major advances in our understanding of molecular biology.

Even though there is much genetic variability in human populations we know there must, nevertheless, be a great deal of scope for environmental effects. The fact that a character is genetic does not by any means imply that it cannot be manipulated by the environment. The cure of PKU by an appropriate diet, and the effect of nutrition on height (which has a very significant genetic component) are simple examples of this. One could even, in principle, conceive of an educational system which compensated for genetic differences in, say, intrinsic ability, by giving less education to the bright and more to the less bright in a way that might truly equalize.

I believe that a common mistake in considering these questions is to imagine that there is a linear scale of quality (IQ in some people's minds) on which high equals good, and low equals bad. The situation is, surely, rather that there simply are differences between people which cannot necessarily be classified as good, bad, high, or low.

There must certainly be implications for education in the existence of genetic differences. Not all will want to learn the same things or learn in the same way, or learn at the same speed. In modern complex societies the need for a variety of individual talents has undoubtedly increased tremendously. Though some educationalists may believe that anybody properly educated can be made proficient in any job, few would surely maintain that the effective cost, however measured, would be the same for each individual. If enough research were to be devoted to genetic differences in response to education, a Utopian goal might suggest that one should identify genotypes more suited to one type of education than another, and then optimize educational procedures to minimize educational costs by matching each type of individual as closely as possible to their best suited occupation. We are still very far from having the sort of knowledge which would allow putting such a programme into practice, perhaps fortunately so. But no educational system will be successful if it ignores the existence of individual variability.

If equality of environment is interpreted literally then, as many have pointed out, in an equal environment all differences that remain will be genetic. It has been suggested that this would lead to a meritocratic society strictly dependent on genetic endowment. Whether this is likely, realistic, or even possible is clearly a moot point, but at the very least we must be aware of the difficulties for the concept of equality in our society given the existence of so much genetic variability.

There are a number of more obvious corrolaries of the existence of genetic variability which follow, for example, from variations in genetic susceptibility to diseases. Mental disease is a particular case in point. Schizophrenia, which occurs with an incidence of one or two per cent in the population, is now generally thought to have a major, if not predominant, genetic component and accounts for a very large fraction of occupied hospital beds. Genetic differences account for the need to match blood for transfusion and transplantation. There are genetically determined allergies and sensitivities to drugs. There are probably genetic factors in alcoholism. Colour blindness is genetically determined and sets some limits to possible occupations. There are almost certainly genetic differences in drug addiction and most probably, also, in tendencies to criminality. All these genetic variations have obvious consequences in our society.

Closely related to the question of individual differences, and often confused with it, is the question of differences between groups of individuals, namely between populations or races. Allegiance to one's own group, be it family, village, country, race, or species seems to me to be so strongly ingrained as to be almost instinctive, and with this allegiance comes its complement racialism. To a biologist a race is just a group of individuals or populations which form some recognizable subdivision of the species. Members of the group share characteristics which, to some extent, distinguish them from members of other groups. Races have traditionally been defined by outwardly obvious features such as skin colour and face shape. Nowadays, however, the only valid approach is to use those individual differences which are simply inherited and which have provided evidence for the genetic uniqueness of the individual. The frequencies of the various genetic types differ from one population to another, and these frequencies can be used to characterize a population. It is important to emphasize that differences between races, as conventionally defined, account for a minority, probably less than 30 per cent, of the total species genetic variability.

Racialism, the belief that some race, usually one's own, is inherently superior and so has a right to dominate, as I have already mentioned, seems to me to follow from a strongly ingrained, almost instinctive, allegiance to one's own group. Many attempts, however, have been made to support racialistic ideas with genetic arguments. The most recent example is the debate on IQ and race differences, notably differences between blacks and whites in the United States, and the question of the extent to which this difference has a genetic component. This is a subject that has been much discussed and I have no time to summarize the arguments here. I should just like to say, however, that many geneticists, including myself, believe that there is no case on present evidence either for assuming, or for not assuming, the existence of a significant genetic component. The data are inadequate and the methodology for answering the question properly is not yet available. Even if there were such an average genetic difference, what relevance would it have to the treatment of the individual, which is the basis for dealing with people in a society free of racial prejudice? The conflict in this example is not created by the scientific knowledge about how we can define population groups using genetic markers. The problem is to use this knowledge for counteracting racialistic tendencies.

There is little doubt that races or populations do seem to have characteristics as a whole which distinguish them in ways that, if these were differences between individuals, one might well think could have a significant genetic component. Yet to a geneticist it is clear that the time available for such differences to arise, especially when taking into account the genetic complexity of the characters that one is usually thinking about, makes it impossible to believe that many, if any, such differences between populations can truly be genetic. Otherwise these characteristics would have to be associated with sufficient biological reproductive advantage to enable the genes determining them to spread quite rapidly through whole populations that are, perhaps, as closely related as the British and the French. Now the cultural spread of a character is quite different from genetic spread. Cultural transmission can be likened to infection, sometimes even to an epidemic. The cultural spread of a trait can occur very rapidly because cultural transmission can take place between people of all ages, whether or not they are related, and is not restricted to movement, according to Mendel's laws, from parent to offspring. Most of the population differences that we normally consider must therefore be cultural, and yet we must try and answer the question as to why they often look as though they might be genetic. I think the answer may be relatively simple. Many cultural changes are initiated by a single individual, or at most a small group of individuals, who are dominant within a society. Once initiated, a trait can be rapidly transmitted culturally. The initiator may have had a particular genetic endowment which enabled him to initiate, but human communication allows others to learn without having the genes needed to initiate. We do not all have to be mathematical geniuses like Newton in order to know and use Newtonian mechanics and the calculus.

My aim in discussing these various examples of applications of advances in genetics to society has been to point out the type of tensions and problems that are raised by the applications of new scientific knowledge. There are, perhaps, at least three important points that arise. First, the proper application of scientific advances will not happen unless the scientist and non-scientist alike consider the problems and deal with them. If applications are left to follow their own course without proper consideration, then scientific advances may easily be misapplied. A second point to emphasize is that the scientist cannot act alone when it comes to applications. He is not necessarily the person that knows best what should be done, and so he must communicate with the non-scientist. Lack of communication may lead to invalid arguments as a basis for action. One small example to my mind, comes from the Law Commission's working paper (No. 47) on the injuries to unborn children. In this reference is made to cases in the 1930s of claims that involvement of a mother in an accident during pregnancy might have led to the resulting child having a club foot, a possibility that seems most unlikely scientifically. The third, and to my mind very important point is the need for good general education in science. This should provide people with an understanding of scientific issues, so that applications create less of a problem simply due to misunderstanding, as in the case of genetic screening programmes. This should also allow all members of our society to help make sensible assessments as to what should be done about applications of scientific advances. In each of these three areas, namely the proper consideration of applications of scientific advances, the need to communicate with the non-scientist, and the need to educate, the scientist has a clear responsibility.

I have purposely not dealt with some of the larger philosophical questions that can be raised, such as: what is science and what is the nature of scientific progress? what is the nature of equality? what is for the good of society? Perhaps some of you will thus accuse me of having been prosaic. I am, however, reminded of the story of a wife, who when asked how she and her husband made their decisions said, 'Oh, that's very easy. He makes all the important decisions, such as what to do about the Middle East War, what to do about the world energy crisis and population increases, and whether to impeach President Nixon. I make all the minor decisions, like where we are going to live, what schools the children will go to, where we are going to have our holidays, and what sort of a car we are going to buy.'

The title of my lecture is 'Biomedical advances—a mixed blessing?' Of course one answer to this question might be to have no advances. But as I emphasized at the beginning, I believe that that cannot be—there is no standing still, only moving forward or backward however slowly, and none of us surely would welcome a new Dark Age. Perhaps, occasionally, a brake has to be applied to some research but that can only, in general, be a minor hiccough in the advance of science.

Thus I believe that even if biomedical advances (and of course other scientific advances) are a mixed blessing—they are a blessing we cannot avoid and it is our job to see to it that the mixture comes out for the best.

4 '… et augebitur scientia'

J. R. RAVETZ
University of Leeds

PERHAPS I should begin by explaining the Latin motto serving as a title for this lecture. It means 'and knowledge shall increase'. It was written by Francis Bacon and is to be found on the frontispiece of his *New Organon* (see Plate). It has several functions in my present discussion. First it advertises my university, Leeds, which has it as its motto. Also, it advertises my field of scholarship, the history and philosophy of science. By scholarly study of that motto in its context, we have recently come to understand Bacon much better than ever before. Finally, I use it as surprise evidence in support of my thesis about progress in science. For the motto's true meaning is not at all a prediction of simple cumulative growth in natural science. But let that wait for a bit; first I should state the thesis which I shall argue in this lecture.

I wish first to show that there is no absolute measure *within* science, for the 'progress of science'. Still less do the fields of application of science have autonomous criteria for progress. Hence we cannot look to natural science, in itself or in its applications, either for an independent measure or for a perfect model for progress in human affairs.

Like all such general propositions, this will strike the various members of this audience in ways depending on their experience and expectations. To some (I hope not too many) it is banal. To others, promising of interest, and finally to still others (again I hope not too many) shocking. This thesis of mine represents a departure from a very important ideology, one of progress in and through science. This ideology is only partly Victorian. It has admirable antecedents deriving from the Enlightenment of the eighteenth century. It still

FRANCISCI
DE VERULAMIO,
Summi Angliæ
CANCELARIJ,
Instauratio
magna.

Multi pertransibunt & augebitur scientia.

Anno

LONDINI
Apud Joannem Billium
Typographum
Regium.

1620.

animates many of the attempts to bring reason to bear on the vexing problems of contemporary life.

It might appear that my thesis, on the relativity of scientific progress, is a part of the 'anti-science' movement which has attracted much notice recently. This is not my intention. I want the thesis to be developed in a positive rather than a negative way. For this, I shall discuss the possibility of an evolution of science as a whole, as distinct from piecemeal progress within it. Indeed, I will consider the argument that such an evolution may be needed for science to play its part in bringing our civilization through its present crisis. What sort of inspirations, or illumination, would be necessary to bring about a proper evolution of science is something on which we can permit ourselves only the most brief of speculations in this lecture. But here too, to complete the surprise offered by my title, Bacon might be our guide.

It might at first seem perverse for anyone to deny that we can correctly identify progress in science; just as it seems wilful ignorance to claim that there is now no genuine progress occurring. Merely to consider the volume of research being done, the increasing stream of publications of all sorts, belies any such doubts. But I shall show that in this respect (as in all others) 'Science' is a complex thing; and here (as elsewhere in human affairs) our judgements become less secure as they become more significant.

The banality of a simple quantitative approach to science is revealed in the smug statement of the 1950s, perhaps now being tactfully forgotten, that 90 per cent of all the world's scientists are now alive. From this we might have deduced that there are now scores of groups of scientists comprising nine men, each one in each group more clever than say, Newton, Archimedes, or Aristotle. In the absence of empirical verification of that deduction, we might have concluded that quantity in science is *not* the same as quality. And as soon as scientists came to recognize the category of 'pointless publication' or what I have called 'shoddy science', and also admitted the existence of immature and ineffective fields of inquiry, *then* it could have been clear that simple quantity of publication guarantees nothing at all in the way of progress. For I have explained how under certain circumstances it is possible for published reports of research to contain no substance, but to consist of unsound data, interpreted by an incoherent argument, which leads to a vacuous conclusion.

This can happen in spite of scientists' good intentions, in an immature field; and it is brought about by a corruption of standards in the case of shoddy science in an established field. Whatever the cause, we are thereby reminded that publication does not equal progress.

Of course there are other criteria of progress; let us pass to them. We may have, in a field of science, a real augmentation of the stock of tested and accepted facts, providing a continuous enrichment and correction of itself. Also, we may have new theories and explanations, not only encompassing new facts and surviving new critical tests but also explaining *why* the earlier theories were inadequate. In this relation between theories, we can locate a genuine assymmetry, from which a unique direction of 'progress' can be inferred quite naturally.

We can therefore be certain that some judgements of progress within science can be meaningful. To the perception of any sensible person, progress can be real. Indeed, so long as we restrict our attention to particular matured, well-defined fields now being actively exploited, the judgement of progress over a small, past time-interval is indubitable. But when we leave the question, 'has this or that field made progress?', for the broader one, 'is science progressing?' the problem changes. In the wider, more significant question, new dimensions of judgement are adjoined. Among these is the implied prediction about the future; the state of 'progressing' lies in the vanishing present. Judgements of significance and value enter, sometimes quite crucially on this and other aspects. I can illustrate these features in the case of a single field.

We might, for example, find an assessment of a field, stating that, in spite of undeniable progress up to the present, it is not progressing well and its future is bleak. This seems paradoxical: progress that is not progressing. How can this be?

First, we must appreciate that the prospects for research in any field can change rapidly. The successful completion of the study of an exciting set of problems can leave a field exhausted, without a clear challenge and stimulus. Or some solid, straightforward progress over the past can be completely devalued by a breakthrough in technique in a neighbouring field. Or even, a record of steady but pedestrian work over a period can finally convince observers that a field is irremediably boring. Any such judgement (and these are only a sample) will serve as a warning to the bright, ambitious men to pass by or even to get out. And in that way the judgement that a field is not progressing well tends to be self-confirming.

We should notice that the judgements of the present state of the field, so crucial for its prospects, are inevitably intuitive and also value-laden. The attributes 'lacking challenge', 'pedestrian', or 'boring' can be argued rationally, with support from public evidence. But they are not amenable to treatment in an 'objective' or 'scientific' style. When judgements of significance go beyond the simple comparison of two closely related established theories, the intuitive component of the judgements increases markedly. Estimates of the future share these features, and are necessarily speculative as well. It is awkward but inevitable that judgements of this character, more appropriate to the conduct of human affairs than to the detailed work of scientific problem-solving, are in fact crucial for the direction of science. The historical, scholarly question concerning the past, 'has there been progress?', is in practice subordinated to the practical, *political* question concerning the future, 'is it progressing?'.

I should remark that even in the less common but apparently simpler case of straight competition between two scientific research programmes (as described by Lakatos), the variety of criteria, and of legitimate estimates concerning the future, ensure that the choice of the 'more progressive' research programme is not straightforward. In the words of Lakatos, there is no 'instant rationality' even in this paradigm ideal case.

I can now state the conclusion of this rather dense little philosophical argument. Judgements of the present and future progress of individual branches of science, which themselves influence the future through their role in decision-making, are made in a style that is 'political' rather than 'scientific', although still fully 'rational'. This thesis is in direct contradiction to the inherited belief that 'science' offers a model for human action that is fundamentally different from, and more certain than, 'politics'.

When we come to consider larger domains within science, we are carried still further from the apparently simple certainties of progress, derived from the comparison of pairs of sets of facts or theories. In the formation of science policy, different fields, each possessing their own criteria of significance, must be assessed for their separate and composite current states and future prospects. This is an urgent task for scientific policy-makers, especially in a period of restricted over-all support. The criteria of quality involved in decision-making at that level can become very broad indeed, to the point of including aspects that we here might consider extraneous to science. Regardless of the sensibilities of any surviving purists, 'progress' now popularly includes such things as a contribution to national prestige and the reduction of inequality of economic opportunity.

Thus science as a whole, including its condition and prospects, is assessed partly by criteria that accommodate many aspects of its involvement in society. Should it be so? Should we not assess science in the large as we do in the small, by purely internal criteria? No. We cannot, even if we want to, for it is impossible to demarcate the section of science which in all respects is to be kept 'pure'. Nor can we correctly claim that such a wider assessment violates the traditions of science. For when we look at history, we find that the justification of science to the wider community has nearly always accepted that the community's values are relevant, although not simply transferable, to the endeavour of science.

By this last argument I have arrived at a rather radical modification of the traditional view of the autonomy of scientific progress. Just previously I showed that in the assessment of science in the small, the application of the criteria of progress must be made in a 'political' rather than 'scientific' style. Now I have shown that for the assessment of science in the large, the criteria themselves are partly derived from the sphere of political life. Although it is not easy for an outsider to make a competent and justifiable criticism of the particular facts achieved by matured natural science, the critic has a right and a duty to examine the currently accepted criteria of progress relevant to science as a whole.

Any such critical examination is likely to encounter a defensive manoeuvre that is fairly commonly employed in debate on established social institutions. This is to place oscillating boundaries around the class of things referred to, by the defining concept in question. Thus for the preservation of the innocence of science, the term is understood to include only that research where *both* the research-worker *and* the supporting agency have the purest of internal criteria of significance. But for the establishment of beneficence, the term 'science' is stretched to include medicine and some technology as well. The elasticity of such moral thinking is well expressed in the critical aphorism, 'Science takes the credit for penicillin, while Society takes the blame for the Bomb.'

Now it is possible to argue that there is a branch of science still so isolated from the world that its 'progress' and its political and moral problems are of a different order from the rest. It could be argued in return that even here there are presuppositions about reality and value imported from the surrounding culture, and then re-exported in increased strength. But this counter-argument can be dangerous unless well handled, so I shall leave it. More to my present point, such ultra-pure science is only a small portion of the total, in significance as well as in size. And it may be that all such pure, positive science is in an inherently unstable position between ideological relevance and sensitivity on the one side (as Darwin's theory of evolution) and industrial and military applicability on the other.

Let us return to that major part of science whose criteria of significance and hence of progress are influenced by political, social, medical, industrial, and military considerations. Here we find several sorts of developments where successive refinements in technique can be described as 'progress' only in an ironic sense. Foremost among these is military R and D, whose influence on all civilian high technology and science is pervasive and incalculable. The institutionalized psychopathology of 'nuclear strategy' is still very much with us, threatening instant annihilation, guaranteeing the impoverishment and distortion of civilian economies, and also corrupting all the sciences and technologies it touches. I blame no man whose conscience sends him to work on such 'dirty science' in the cause of defence of his nation. But how many of our eminent scientists have recently reminded the world that this evil continues still to grow and spawn?

Other fields where the appellation 'progress' seems twisted are in what I call 'runaway technology'. A supersonic air transport promises to be quicker than subsonic for its passengers, so therefore it must be better for all. As to sonic booms and similar irritations, the class of those people who are bothered by them can be officially considered as 'statistically negligible', on the basis of a long series of avoidances of genuine tests of nuisance. There have been a few scientists and engineers involved in the campaign against this monstrous farce; they have given a lot of help to the schoolmasters and journalists who have made the campaign effective. But the silent acquiescence of the British scientific and technological Establishment in supersonic 'progress' is not a recommendation for their honour or even for their perspicacity.

In the face of such examples of 'progress in technology' (and there are many), one must agree with Ivan Illich that the pattern of evolution of tools includes 'watersheds' ending the phase of benefit and introducing that of ever-increasing human cost.

Again, one may criticize 'reckless science', as of those biomedical engineers who fully discharge their social responsibilities by warning society that it has just ten years in which to prepare for the consequences of their own noble and disinterested search for Truth. I here spell the word with a capital T, so as to ensure its morally absolute character.

Please do not conclude that I consider all or even a majority of scientists to be personally tainted by association with the pathological situations I have described. Not at all; science as a whole is still, I think, on the healthy side of the watershed I mentioned. Moreover, it is only a defective social system that fails to operate automatically, by the efforts of completely honourable men. And up to now at least, the world of science has run well, its wider criteria of progress generally in harmony with those of both rulers and ruled in our society. But the pathologies of scientific 'progress', that I have indicated, can be described and debated only because that automatic mechanism, that nearly universal consensus on the nature of 'progress', now works not quite so well. Indeed, we may now be entering a phase of deep crisis for the most basic idea of 'progress' that has animated society and science alike in recent generations. For the 'limits to growth' entail that we can no longer buy our way out of the sin of poverty. We can no longer offer the hope that poor men will some day live as greedily as the rich. Henceforth 'progress' and 'growth' must be of the spirit.

From this new historical situation, I conclude that our idea of 'progress' will need to change. But this itself need not cause consternation; the ideas of 'progress' that guide science have been in continuous evolution. To illustrate this, I shall consider the pioneers of the Scientific Revolution of the seventeenth century, in particular Francis Bacon. In this connection I can introduce a surprise. Let us now consider the motto of my title: *et augebitur scientia*. This seems to express admirably the inductive, fact-gathering aspect of Bacon's philosophy that was so admired by his Victorian interpreters. As it happens, these three words are only the second half of the motto. They are preceded by the pair: *multi pertransibunt* . . . Now, this extra bit can be explained, especially in the context of the frontispiece where the full motto appears. It is translated, 'many shall go to and fro'. For explanation of the reference to travelling, we have Bacon's own mention of the 'intellectual' sphere, analogous to the material one, and his proposals for voyages of discovery there, to match those of the earlier explorers in their little wooden sailing boats.

This interpretation of the motto might be acceptable if Bacon were only a scientist, who had found somewhere a nice Latin epigram for his book. But he was a philosopher and a prophet. Also, he was a master of humanistic learning, and a great craftsman of prose, Latin and English. So we look for the source of the motto, and we find it in the Bible: Daniel 12:4. What is it doing there? It is describing The Last Days, albeit in a somewhat obscure fashion. Did Bacon realize this? Of course he did: he quoted the motto in Aphorism 93 of Book 1 of the New Organon, using its prophetic message for his own purpose. We read:

Nor should the prophecy of Daniel be forgotten, touching the last ages of the World:—'Many shall go to and fro, and knowledge shall be increased:' clearly intimating that the thorough passage of the world (which now by so many distant voyages seems to be accomplished, or in the course of accomplishment), and the advancement of the sciences, are destined by fate, that is, by Divine Providence, to meet in the same age.

In using Scripture so literally as a guide to human affairs, Bacon was not being simple-minded, reactionary, or eccentric. For him, and for generations after him, human history and human progress were conceived in Biblical terms. The functions that in our contemporary world have been performed variously by 'the classless society', the 'world co-operative common-wealth', or simply 'growth' were then an affair of the Millennium. This formulation of the goal of human striving provided content, guidelines for achievement, and perhaps timing as well.

Bacon saw real practical significance in the progress, in his own time and earlier, in voyages of discovery and in the increase of knowledge. They were good evidence for him that the 'last days' were truly at hand; that the Father of Lights would soon see to the restoration of mankind to felicity. The progress of the sciences, in which knowledge and power were to meet in one, was nothing more nor less than the central task of this sacred and urgent work of bringing about the Millennium.

The pursuit of the Millennium is an occupation that unfortunately does not guarantee the enhancement of man's gentler qualities. Happily, Bacon's insights were in close sympathy with a tradition we can call 'philanthropic' science, a combination of alchemy, mysticism, and social reform. In the first book of the New Organon, where he offered his concentrated wisdom on the sciences and their improvement, it is hard to find a passage where science is discussed in abstraction from ethics and morality. It is plain that the 'Instauration' of the sciences, while contributing to the great Reform of man and religion, also requires and fosters a reform among its practitioners. In the unpublished writings of his middle years, Bacon made it plain that he envisaged a fraternity of purified souls, as God's vehicle for accomplishing this good work. There were many such imagined organizations at that time, when the world approached the centenary of Luther's Reformation. Bacon's probably had less real existence than most, for there is no evidence of any colleagues of his who gave more than friendly sympathy to his projects. His was a lonely path.

Appreciation of Bacon's millennarian perspective and strategy helps us to understand many features of the style and doctrine of his writings. His hints of an esoteric doctrine are explained in terms of the true knowledge being restricted to those sufficiently pure to receive it. Also, his assurance that the great work begins softly, and noiselessly, reflects more a private society than a Royally endorsed academy. And in his text, the Latin term *res*, capable of being translated as 'fact' or indeed 'case' in the legal sense, can also mean *thing*—the real thing that Adam named and which we can name again.

Bacon's more formal teaching includes methodology, epistemology, and ethics. The first is the most explicit, taking up all of Book II of the New Organon; it has been laboured and belaboured by generations of commentators. His developed epistemology, is, oddly enough, purely negative. His doctrine of the Idols provides a very powerful analysis of error through a genetic approach (a sort of antiépistémologie génétique). In sequence we have The Tribe, or the universal weakness of our equipment for learning and knowing; and The Cave, our individual idiosyncracies. Then the Market Place, where we deal in the clipped coinage of vulgarized ideas as purveyed by parents and teachers; and finally the Theatre, where Professors strut and spout. This is for me personally a more compelling scepticism than Descartes' overheated Meditations, because it has the flavour of real life with its real disillusions. Descartes' ladder out of his pit of unknowing was logic; Bacon's was ethics; and to this I now turn. I believe that he still speaks to us today.

There is no single passage where Bacon exhibits an ethical system, so what I say here must be an imaginative reconstruction, particularly in my assignment of phases. First we have the stern, Puritanical injunctions to discipline, severity, and even chastity. There is a quite explicit analogy drawn between vicious and virtuous scientific work on the one hand, and lascivious sex and good old boring English conjugal relations on the other. This is, perhaps, Bacon's first reaction to the overwhelming impression of levity and vanity in intellectual pursuits around him. Indeed, he inveighs like a prophet of old against the prideful sins of mankind, the learned as well as the vulgar, and their baneful consequences. This moralizing fits well with a narrow conception of Bacon's inductive methodology. It also seems to anticipate the recent social patterns that tend to make scientists into narrow 'convergers' rather than poetic 'divergers'.

But Bacon does not stop there. A positive ethical teaching emerges, in his scientific writings and also in his private meditations. It starts with charity, not noble sentiments but particular, practical actions for relieving the suffering of particular people. He observes that all miracles wrought by Christ were for direct human benefit, none for showmanship or harm. His finest invective is reserved for hypocrites:

The ostentation of hypocrites is ever confined to the works of the first table of the law, which prescribes our duties to God. The reason is twofold: both because works of this class have a greater pomp of sanctity, and because they interfere less with their desires. The way to convict a hypocrite, therefore, is to send him from the works of sacrifice to the works of mercy.

His conception of charity is revealed by his describing its 'summit and exaltation'; 'thus if evil overtake your enemy from elsewhere, and you in the inmost recesses of your heart are grieved and distressed, and feel no touch of joy, as thinking that the day of your revenge and redress has come . . . .'

And the role of charity in the reform of the sciences is seen in his prayers:

Lastly, I would address one general admonition to all; that they consider what are the true ends of knowledge, and that they seek it not either for pleasure of mind, or for contention, or for superiority to others, or for profit, or fame, or power, or any of these inferior things; but for the benefit and use of life; and that they perfect and govern it in charity. For it was from lust of power that the angels fell, from lust of knowledge that men fell; but of charity there can be no excess, neither did angel or man ever come in danger by it.

There is yet something else. In my reconstruction, with discipline as preparation, and charity as the way, there lies something at the end: innocence. Of this there are only a few hints. One is a promise, concerning the possibility of real knowledge:

> Whereas of the sciences which regard nature, the divine philosopher declares that 'it is the glory of God to conceal a thing, but it is the glory of the King to find a thing out'. Even as though the divine nature took pleasure in the innocent and kindly sport of children playing at hide and seek, and vouch-safed of his kindness and goodness to admit the human spirit for his playfellow at that game.

This is not a mere figure of speech; for innocence has an important function in Bacon's epistemology. It is, in fact, the *only* cure for the infirmities described in the Four Idols. There we read in Aphorism 68:

> So much concerning the several classes of Idols, and their equipage: all of which must be renounced and put away with a fixed and solemn determination, and the understanding thoroughly freed and cleansed; the entrance into the kingdom of man, founded on the sciences, being not too much other than the entrance into the kingdom of heaven, where-into none may enter except as a little child.

Bacon's ethical ideas were not dominant in the Scientific Revolution. Indeed, they were more in sympathy with the alchemical and pietistic traditions that rejected hard atomism and so were eventually pushed off the scientific scene. At the opposite extreme was the tradition personified in Galileo, with its roots in civil and military engineering. This was more effective in solving the problems it set, and also less demanding in its ethical aspects. Midway between the two was Descartes, metaphysician and geometer but also a man of spiritual experience. For he put medicine as the primary goal of natural science, and he also uttered a 'Scientist's Oath' of devastating simplicity. At the end of the *Discours* he said, 'I could not work on projects that are useful to some only by being harmful to others.' (I am reminded by Sir Isaiah Berlin that Bacon was far from pacifism in his recommendations on contemporary affairs. This is likely to be an aspect of the unresolved contradiction between his humanistic aspirations and his responsibilities to his Sovereign.) By that particular criterion, science has had a lot of anti-progress since then. And in the measure that natural science has not succeeded in reforming men's souls, it has in Bacon's perspective, been a failure and a corruption.

I have given brief sketches of two approaches to the problem of progress in science, one philosophical and the other historical. I hope that they suffice for my present conclusion, and that it is not necessary for me to give an exhaustive catalogue of criteria of progress in use now or in the past. What we see is that the 'progress' of science is assessed and perceived within frameworks of ideas that are historically and culturally conditioned. Whether there is a *noumenon* of real but unreachable scientific progress, behind these phenomena, is a Kantian problem that I hope we can leave to one side for the present.

It can be argued that, in such unstable times as these, our need for criteria of progress is even more urgent than before. Less can be left to routine, or to chance: more must be subjected to conscious decision and choice. The argument is correct; the only question is the source of these criteria. In an old and honourable tradition, they have been thought to originate within natural science and then to irradiate out to civilization in general. However justified this view was in its time, I believe I have shown that it is no longer tenable: we now know that 'science' is too various, changeable and ill-defined to serve as a permanent bench-mark for civilization.

It would certainly be a betrayal of science and much else, if the basic values by which its progress is assessed, were to be picked up from each passing fashion in politics. As science is an integral part of our civilization, it takes its values from that totality, and in return contributes to its development. The values which predominate at any time will form a family, always changing somewhat and not completely consistent among themselves. Hence styles in science can have a legitimate and stimulating diversity. But certain particular values are essential to science as we understand it; and if these are neglected then any further 'progress' will be illusory and short-lived. Among these I believe we can identify curiosity, and also intellectual integrity. I would go further and argue that in the further development of science the factual content is less important than the preservation of these defining values. Might I remark that here, as on previous occasions, I find myself laying down my own path in order to arrive at substantial agreement with what was said long ago by Sir Karl Popper.

Having said that much in agreement with tradition, I must go on to comment on the insufficiency of that tradition. The scientific research-worker today, imbued with the appropriate sort of puzzle-solving curiosity and professional honesty, comes nowhere near to being a realization of Bacon's ideal of the man of true learning. Similarly, the many material benefits achieved by the applications of science, while raising men (poor and rich alike) towards a civilized stage, still as yet leave radical flaws untouched. Until quite recently it was plausible to say that more of the same traditional science would overcome the remaining obstacles. But now we can speak of a crisis in science, a special and acute focus of a crisis in our civilization.

I have already indicated some of the pathological states of science, as dirty science, runaway technology, and reckless science. How are these to be contained and controlled? We cannot simply leave it to those eminent men who have gained great wisdom through a career of scientific research and who now lead the scientific community. For these abuses, and others, have developed while these same eminent men, or their immediate predecessors, have been presiding over the world of science. If the leaders of science and their followers are an integral part of the problem, can they so simply constitute the whole solution? For this reason I believe it is justifiable and necessary for some of us to think critically about the social and ethical aspects of scientific progress. I also believe that Bacon's ethical critique and insights still apply to the present, and I hope to use his work for further illumination.

In my own writings I have indicated a new style of science, called 'critical science', in which the traditional virtues of curiosity and honesty would be enriched by another: commitment to humanity. In spite of the inevitably disorganized and fragmented state of 'critical science' some necessary features of its style are becoming discernible. For me they indicate a hopeful path of progress for science. Moreover, the conditions for the maturing of critical science, necessary and I hope sufficient, give some hints to a direction of the future evolution of science.

To appreciate these possible changes we should recall how stylized are our own systems of production, communication, and control in science. The basic unit is an atom, the research paper: a self-contained product of the creative work of a single identifiable person (or small team). These atoms have links of course, those of content being in fair correspondence with social links between the producers. These social groupings are in a tension, frequently creative, between the one function of fostering innovations and the other one of protecting the property (realized in the viability of past work) of its members. This tension is accommodated partly by fairly tight restrictions on the content and on the style of approved or permitted productions of members. To venture into other fields is to risk disapproval abroad and at home; and to dabble in political matters is to court condemnation as 'unsound'.

The presence of such a social system of science now ensures a powerful filter in respect of commitment for those entering 'critical science'. There may be zealots there, but as yet no time-servers. For the real problems thrown at us by the blunders of the techno-machine do not respect disciplines, nor are they amenable to the isolated, atomic style of work. They engage whole men, demandingly and hazardously. Also, they foster conceptions of Nature and of knowledge that have been rather suppressed in recent times. These involve thinking in wholes; respecting non-quantifiable attributes, including human values, as real; and seeing oneself and one's problem in a living, historically developing context. Some established natural sciences already enjoy such a style, but it must be confessed that they are a minority.

I hope that the style of research I call 'critical' will soon get another name, as its usefulness and promise appeals to the imagination of scientists, young and old. Then we will be able to engage in something far more important than cumulative progress within the sciences: the evolution of science itself. This must involve all three elements that constitute the activity of science: objects of inquiry, methods of research, and functions of achieved results. In brief: metaphysics, methods and (broadly) ethics. How will those change?

At this point I pass to a personal concluding speculation, which I hope you will find worthwhile. Francis Bacon lived through a watershed, when in the consciousness of most educated Europeans, the natural world was on the point of becoming disenchanted and dehumanized. The ancient sciences that presupposed meaning in the cosmos, as alchemy and astrology, were on the point of being discredited. Bacon's friend and pupil Hobbes produced a metaphysics which (like that of Descartes) reflected the desperate disillusion of the early seventeenth century. Hobbes' universe, natural and social, was of atoms; life is nothing but motion. Certainly in his universe there is no consciousness, intelligence, or power except in aggregates of tangible atoms. We have lived with this metaphysics for a long time, built it into many of our social arrangements, and in its terms conquered many other peoples and many aspects of matter. I need not remark on the complexities and contradictions within this historical process and the modifications of 'possessive individualism' in recent times. Hence it is a matter of personal judgement for me to say that now, after some three centuries, Hobbist atomism in thought and life, the billiard-ball universe of the anti-religion of Science, is outmoded and dysfunctional.

A new complex of metaphysics, methods and ethics for science, is scarcely born. We do not know its shape, still less can we be sure of the way to its achievement. Yet if science is to evolve, seeing civilization through and beyond its present crisis, it will need to be in some such direction away from its atomistic ideology. The wisdom and insight of Bacon can be our guide in this new journey. What I have called discipline, charity, and innocence could equally well be described as humility, love, and bliss. What would be the outlines of a new science in Bacon's vision? First without the need for the protection of property, the bolstering of lonely egos, all the fields, disciplines, subjects, departments, divisions, all that scaffolding that now defaces and clutters the edifice of knowledge, could be erected lightly and dismantled quickly on need. Not fearing ridicule from a strait-laced colleague community, scientists could openly be poets, fully creative in every way. And the domains of investigation could include all that human benefit and reverent curiosity call forth: the natural, social, and spiritual worlds blended in our understanding as they are in fact.

Will this frankly Utopian state of affairs be brought about by the same old style of material progress and social reform? The insufficiency of the first of these is already becoming apparent through the absolute limits of our planet to support our material exploitation. Also I fear that we shall need to wait a long time for the withering away of the State of its own accord. So perhaps something else can be injected into this new process of Reform, a new purification of men and knowledge. According to the scientific common sense of Hobbes, there can be nothing else. But Bacon almost certainly would have disagreed. For him the Father of Lights was present, ready to forgive mankind his errors and to restore him to grace. Bacon pursued his multiple career of service to his nation and his God, burdened by frustration, disappointment, and tragedy, with a constant serenity. He had, I am almost certain, glimpses of a Knowledge and a pure self-effulgent Light that gave him direction, certainty, and inner peace. Much to the detriment of science, all too few scientists have had the inner vision of this Light. But it was still a commonplace in the century of the Scientific Revolution, illuminating men as diverse as Henry Vaughan the poet and George Fox the Quaker. Should we all be granted the experience of this Light to guide the evolution of science, then Bacon's vision and endeavour will be justified, and *true* knowledge will increase.

5 *The steep and thorny way to a science of behaviour*

B. F. SKINNER

Harvard University

A CRITIC contends that a recent book of mine does not contain anything new, that much the same thing was said more than four centuries ago in theological terms by John Calvin. You will not be surprised, then, to find me commending to you the steep and thorny way to that heaven promised by a science of behaviour. But I am not one of those ungracious pastors, of whom Ophelia complained, who 'recking not their own rede themselves tread the primrose path of dalliance'. No, I shall rail at dalliance, and in a manner worthy, I hope, of my distinguished predecessor. If I do not thunder or fulminate, it is only because we moderns can more easily portray a truly frightening hell. I shall merely allude to the carcinogenic fallout of a nuclear holocaust. And no Calvin ever had better reason to fear his hell, for I am proceeding on the assumption that nothing less than a vast improvement in our understanding of human behaviour will prevent the destruction of our way of life or of mankind.

Why has it been so difficult to be scientific about human behaviour? Why have methods which have been so prodigiously successful almost everywhere else failed so ignominiously in this one field? Is it because human behaviour presents unusual obstacles to a science? No doubt it does, but I think we are beginning to see how they may be overcome. The problem, I submit, is digression. We have been drawn off the straight and narrow path, and the word *diversion* serves me well by suggesting not only digression but dalliance. In this lecture I shall analyse some of the diversions peculiar to the field of human behaviour which seem to have delayed our advance towards the better understanding we desperately need.

I must begin by saying what I take a science of behaviour to be. It is, I assume, part of biology. The organism that behaves is the organism that breaths, digests, conceives, gestates, and so on. As such, it will eventually be described and explained by the anatomist and physiologist. So far as behaviour is concerned, they will give us an account of the genetic endowment of the species and tell us how that endowment changes during the lifetime of the individual and why, as a result, the individual then responds in a given way upon a given occasion. Despite remarkable progress, we are still a long way from a satisfactory account in such terms. We know something about the chemical and electrical effects of the nervous system and the location of many of its functions, but the events which actually underlie a single instance of behaviour—as a pigeon picks up a stick to build a nest, or a child a block to complete a tower, or a scientist a pen to write a paper—are still far out of reach.

Fortunately, we need not wait for further progress of that sort. We can analyse a given instance of behaviour in its relation to the current setting and to antecedent events in the history of the species and the individual. Thus, we do not need an explicit account of the anatomy and physiology of genetic endowment in order to describe the behaviour, or the behavioural processes, characteristic of a species, or to speculate about the contingencies of survival under which they might have evolved, as the ethologists have convincingly demonstrated. Nor do we need to consider anatomy and physiology in order to see how the behaviour of the individual is changed by his exposure to contingencies of reinforcement during his lifetime and how as a result he behaves in a given way on a given occasion. I must confess to a predilection here for my own speciality, the experimental analysis of behaviour, which is a quite explicit investigation of the effects upon individual organisms of extremely complex and subtle contingencies of reinforcement.

There will be certain temporal gaps in such an analysis. The behaviour and the conditions of which it is a function do not occur in close temporal or spatial proximity, and we must wait for physiology to make the connection. When it does so, it will not invalidate the behavioural account (indeed, its assignment could be said to be specified by that account), nor will it make its terms and principles any the less useful. A science of behaviour will be needed for both theoretical and practical purposes even when the behaving organism is fully understood at another level, just as much of chemistry remains useful even though a detailed account of a single instance may be given at the level of molecular or atomic forces. Such, then, is the science of behaviour from which I suggest we have been diverted—by several kinds of dalliance to which I now turn.

Very little biology is handicapped by the fact that the biologist is himself a specimen of the thing he is studying, but that part of the science with which we are here concerned has not been so fortunate. We seem to have a kind of inside information about our behaviour. It may be true that the environment shapes and controls our behaviour as it shapes and controls the behaviour of other species—but *we* have feelings about it. And what a diversion they have proved to be! Our loves, our fears, our feelings about war, crime, poverty, and God—these are all basic, if not ultimate, concerns. And we are as much concerned about the feelings of others. Many of the great themes of mythology have been about feelings—of the victim on his way to sacrifice or of the warrior going forth to battle. We read what poets tell us about their feelings, and we share the feelings of characters in plays and novels. We follow regimens and take drugs to alter our feelings. We become sophisticated about them in, say, the manner of La Rochefoucauld, noting that jealousy thrives on doubt, or that the clemency of a ruler is a mixture of vanity, laziness, and fear. And with some psychiatrists we may even try to establish an independent science of feelings in the intrapsychic life of the mind or personality.

And do feelings not have some bearing on our formulation of a science of behaviour? Do we not strike because we are angry and play music because we feel like listening? And if so, are our feelings not to be added to those antecedent events of which behaviour is a function? This is not the place to answer such questions in detail, but I must at least suggest the kind of answer that may be given. William James questioned the causal order: perhaps we do not strike because we are angry but feel angry because we strike. That does not bring us back to the environment, however, although James and others were on the right track. What we feel are conditions of our bodies, most of them closely associated with behaviour and with the circumstances in which we behave. We both strike *and* feel angry for a common reason, and that reason lies in the environment. In short, the bodily conditions we feel are *collateral products* of our genetic and environmental histories. They have no explanatory force; they are simply additional facts to be taken into account.

Feelings enjoy an enormous advantage over genetic and environmental histories. They are warm, salient, and demanding, where facts about the environment are easily overlooked. Moreover, they are *immediately* related to behaviour, being collateral products of the same causes, and have therefore commanded more attention than the causes themselves, which are often rather remote. In doing so, they have proved to be one of the most fascinating attractions along the path of dalliance.

A much more important diversion has for more than 2000 years made any move towards a science of behaviour particularly difficult. The environment acts upon an organism at the surface of its body, but when the body is our own, we seem to observe its progress beyond that point—for example, we seem to see the real world become experience, a physical presentation become a sensation or a percept. Indeed, this second stage may be all we see. Reality may be merely an inference, and according to some authorities a bad one. What is important may not be the physical world on the far side of the skin, but what that world means to us on this side.

Not only do we seem to see the environment on its way in, we seem to see behaviour on its way out. We observe certain early stages— wishes, intentions, ideas, and acts of will—before they have, as we say, found expression in behaviour. And as for our environmental history, that can also be viewed and reviewed inside the skin for we have tucked it all away in the storehouse of our memory. Again this is not the place to present an alternative account, but several points need to be made. The behaviouristic objection is not primarily to the metaphysical nature of mind stuff. I welcome the view, clearly gaining in favour among psychologists and physiologists and by no means a stranger to philosophy, that what we introspectively observe, as well as feel, are states of our bodies. But I am not willing to give introspection much of a toehold even so, for there are two important reasons why we do not discriminate precisely among our feelings and states of mind and hence why there are many different philosophies and psychologies.

In the first place, the world within the skin is private. Only the person whose skin it is can make certain kinds of contact with it. We might expect that the resulting intimacy should make for greater clarity, but there is a difficulty. The privacy interferes with the very process of coming to know. The verbal community which teaches us to make distinctions among things in the world around us lacks the information it needs to teach us to distinguish events in our private world. It cannot teach us the difference between diffidence and embarrassment, for example, as readily or as accurately as that between red and blue or sweet and sour.

Secondly, the self-observation which leads to introspective knowledge is limited by anatomy. It arose very late in the evolution of the species, because it is only when a person begins to be asked about his behaviour and about why he behaves as he does that he becomes conscious of himself in this sense. Self-knowledge depends upon language and in fact upon language of a rather advanced kind, but when questions of this sort first began to be asked, the only nervous systems available in answering them were those which had evolved for entirely different reasons. They had proved useful in the internal economy of the organism, in the co-ordination of movement, and in operating upon the environment, but there was no reason why they should be suitable in supplying information about those very extensive systems which mediate behaviour. To put it crudely, introspection cannot be very relevant or comprehensive because the human organism does not have nerves going to the right places.

One other problem concerns the nature and location of the knower. The organism itself lies, so to speak, between the environment that acts upon it and the environment it acts upon, but what lies between those inner stages—between, for example, experience and will? From what vantage point do we watch stimuli on their way into the storehouse of memory or behaviour on its way out to physical expression? The observing agent, the knower, seems to contract to something very small in the middle of things.

In the formulation of a science with which I began, it is the *organism as a whole* that behaves. It acts in and upon a physical world, and it can be induced by a verbal environment to respond to some of its own activities. The events observed as the life of the mind, like feelings, are *collateral products*, which have been made the basis of many elaborate metaphors. The philosopher at his desk asking himself what he really knows, about himself or the world, will quite naturally begin with his experiences, his acts of will, and his memory, but the effort to understand the mind from that vantage point, beginning with Plato's supposed discovery, has been one of the great diversions which have delayed an analysis of the role of the environment.

It did not, of course, take inside information to induce people to direct their attention to what is going on inside the behaving organism. We almost instinctively look inside a system to see how it works. We do this with clocks, as with living systems. It is standard practice in much of biology. Some early efforts to understand and explain behaviour in this way have been described by Onians in his classic *Origins of European thought*.[1] It must have been the slaughterhouse and the battlefield which gave man his first knowledge of anatomy and physiology. The various functions assigned to parts of the organism were not usually those which had been introspectively observed. If Onians is right, the *phrénes* were the lungs, intimately associated with breathing and hence, so the Greeks said, with thought and, of course, with life and death. The *phrénes* were the seat of *thumós*, a vital principle whose nature is not now clearly understood, and possibly of ideas, in the active sense of Homeric Greek. (By the time an idea had become an object of quiet contemplation, interest seems to have been lost in its location.) Later, the various fluids of the body, the humours, were associated with dispositions, and the eye and the ear with sense data. I like to imagine the consternation of that pioneer who first analysed the optics of the eyeball and realized that the image on the retina was upside down!

Observation of a behaving system from without began in earnest with the discovery of reflexes, but the reflex arc was not only not the seat of mental action, it was taken to be a usurper, the spinal reflexes replacing the *Rückenmarkseele* or soul of the spinal cord, for example. The reflex arc was essentially an anatomical concept, and the physiology remained largely imaginary for a long time. Many years ago I suggested that the letters CNS could be said to stand, not for the central nervous system, but for the conceptual nervous system. I had in mind the great physiologists Sir Charles Sherrington and Ivan Petrovich Pavlov. In his epoch-making *Integrative action of the nervous system*[2] Sherrington had analysed the role of the synapse, listing perhaps a dozen characteristic properties. I pointed out that he had never seen a synapse in action and that all the properties assigned to it were inferred from the behaviour of his preparations. Pavlov had offered his researches as evidence of the activities of the cerebral cortex though he had never observed the cortex in action but had merely inferred its processes from the behaviour of his experimental animals. But Sherrington, Pavlov, and many others were moving in the direction of an instrumental approach, and the physiologist is now, of course, studying the nervous system directly.

The conceptual nervous system has been taken over by other disciplines—by information theory, cybernetics, systems analyses, mathematical models, and cognitive psychology. The hypothetical structures they describe do not depend upon confirmation by direct observation of the nervous system for that lies too far in the future to be of interest. They are to be justified by their internal consistency and the successful prediction of selected facts, presumably not the facts from which the constructions were inferred.

These disciplines are concerned with how the brain or the mind must work if the human organism is to behave as it does. They offer a sort of thermodynamics of behaviour without reference to molecular action. The computer with its apparent simulation of Thinking Man supplies the dominant analogy. It is not a question of the physiology of the computer—how it is wired or what type of storage it uses—but of its behavioural characteristics. A computer takes in information as an organism receives stimuli, and processes it according to an inbuilt program as an organism is said to do according to its genetic endowment. It encodes the information, converting it to a form it can handle, as the organism converts visual, auditory, and other stimuli into nerve impulses. Like its human analogue it stores the encoded information in a memory, tagged to facilitate retrieval. It uses what it has stored to process information as received, as a person is said to use prior experience to interpret incoming stimuli, and later to perform various operations—in short, to compute. Finally, it makes decisions and behaves: it prints out.

There is nothing new about any of this. The same things were done thousands of years ago with clay tiles. The overseer or tax collector kept a record of bags of grain, the number, quality, and kind being appropriately marked. The tiles were stored in lots as marked; additional tiles were grouped appropriately; the records were eventually retrieved and computations made; and a summary account was issued. The machine is much swifter, and it is so constructed that human participation is needed only before and after the operation. The speed is a clear advantage, but the apparent autonomy has caused trouble. It has seemed to mean that the mode of operation of a computer resembles that of a person. People do make physical records which they store and retrieve and use in solving problems, but it does not follow that they do anything of the sort in the mind. If there were some exclusively subjective achievement, the argument for the so-called higher mental processes would be stronger, but, so far as I know, none has been demonstrated. True, we say that the mathematician sometimes intuitively solves a problem and only later, if at all, reduces it to the steps of a proof, and in doing so he seems to differ greatly from those who proceed step by step, but the differences could well be in the evidence of what has happened, and it would not be very satisfactory to define thought simply as un-explained behaviour.

Again, it would be foolish of me to try to develop an alternative account in the time available. What I have said about the intro-spectively observed mind applies as well to the mind that is con-structed from observations of the behaviour of others. The *accessi-bility* of stored memories, for example, can be interpreted as the *probability* of acquired behaviours, with no loss in the adequacy of the treatment of the facts, and with a very considerable gain in the assimilation of this difficult field with other parts of human behaviour.

I have said that much of biology looks inside a living system for an explanation of how it works. But not all of biology. Sir Charles Bell could write a book on the hand as evidence of design. The hand was evidence; the design lay elsewhere. Darwin found the design, too, but in a different place. He could catalogue the creatures he discovered on the voyage of the *Beagle* in terms of their form or structure, and he could classify barnacles for years in the same way, but he looked beyond structure for the principle of natural selection. It was *the relation of the organism to the environment* that mattered in evolution. And it is the relation to environment which is of pri-mary concern in the analysis of behaviour. Hence, it is not enough to confine oneself to organization or structure, even of the most pene-trating kind. That is the mistake of most of phenomenology, exis-tentialism, and the structuralism of anthropology and linguistics. When the important thing is a relation to the environment, as in the phylogeny and ontogeny of behaviour, the fascination with an inner system becomes a simple digression.

We have not advanced more rapidly to the methods and instru-ments needed in the study of behaviour precisely because of the diverting preoccupation with a supposed or real inner life. It is true that the introspective psychologist and the model builder have investigated environments, but they have done so only to throw some light on the internal events in which they are interested. They are no doubt well-intentioned helpmates, but they have often simply misled those who undertake the study of the organism as a behaving system in its own right. Even when helpful, an observed or hypothetical inner determiner is no explanation of behaviour until it has itself been explained, and the fascination with an inner life has allayed curiosity about the further steps to be taken.

I can hear my critics: 'Do you really mean to say that all those who have inquired into the human mind, from Plato and Aristotle through the Romans and scholastics, to Bacon and Hobbes, to Locke and the other British empiricists, to John Stuart Mill, and to all those who began to call themselves psychologists—that they have all been wasting their time?' Well, not all their time, fortunately. Forget their purely psychological speculations, and they were still remarkable people. They would have been even more remarkable, in my opinion, if they could have forgotten that speculation themselves. They were careful observers of human behaviour, but the intuitive wisdom they acquired from their contact with real people was flawed by their theories.

It is easier to make the point in the field of medicine. Until the present century very little was known about bodily processes in health and disease from which useful therapeutic practices could be derived. Yet it should have been worthwhile to call in a physician. Physicians saw many ill people and should have acquired a kind of wisdom—unanalysed perhaps, but still of value in prescribing simple treatments. The history of medicine, however, is largely the history of barbaric practices—blood-lettings, cuppings, poultices, purgations, violent emetics—which much of the time must have been harmful. My point is that these measures were not suggested by the intuitive wisdom acquired from familiarity with illness; they were suggested by *theories*, theories about what was going on inside an ill person. Theories of the mind have had a similar effect—less dramatic, perhaps, but quite possibly far more damaging. The men I have mentioned made important contributions in government, religion, ethics, economics, and many other fields. They could do so with an intuitive wisdom acquired from experience. But philosophy and psychology have had their bleedings, cuppings, and purgations too, and they have obscured simple wisdom. They have diverted wise people from a path which would have led more directly to an eventual science of behaviour. Plato would have made far more progress towards the good life if he could have forgotten those shadows on the wall of his cave.

Still another kind of concern for the self distracts us from the programme I have outlined. It has to do with the individual, not as an object of self-knowledge, but as an agent, an initiator, a creator. I have developed this theme in *Beyond freedom and dignity*.[3] We are more likely to give a person credit for what he does if it is not obvious that it can be attributed to his physical or social environment, and we are likely to feel that truly great achievements must be inexplicable. The more derivative a work of art, the less creative; the more conspicuous the personal gain, the less heroic an act of sacrifice. To obey a well-enforced law is not to show civic virtue. We see a concern for the aggrandizement of the individual, for the maximizing of credit due him, in the self-actualization of so-called humanistic psychology, in some versions of existentialism, in Eastern mysticism and certain forms of Christian mysticism in which a person is taught to reject the world in order to free himself for union with a divine principle or with God, as well as in the simple structuralism which looks to the organization of behaviour rather than to the antecedent events responsible for that organization. The difficulty is that, if the credit due a person is infringed by evidences of the conditions of which his behaviour is a function, then a scientific analysis appears to be an attack on human worth or dignity. Its task is to explain the hitherto inexplicable and hence to reduce any supposed inner contribution which has served in lieu of explanation. Freud moved in this direction in explaining creative art, and it is no longer just the cynic who traces heroism and martyrdom to powerful indoctrination. The culminating achievement of the human species has been said to be the evolution of man as a moral animal, but a simpler view is that it has been the evolution of cultures in which people behave morally although they have undergone no inner change of character.

Even more traumatic has been the supposed attack on freedom. Historically, the struggle for freedom has been an escape from physical restraint and from behavioural restraints exerted through punishment and exploitative measures of other kinds. The individual has been freed from features of his environment arranged by governmental and religious agencies and by those who possess great wealth.

The success of that struggle, though it is not yet complete, is one of man's great achievements, and no sensible person would challenge it. Unfortunately, one of its by-products has been the slogan that 'all control of human behaviour is wrong and must be resisted'. Nothing in the circumstances under which man has struggled for freedom justifies this extension of the attack on controlling measures, and we should have to abandon all the advantages of a well-developed culture if we were to relinquish all practices involving the control of human behaviour. Yet new techniques in education, psychotherapy, incentive systems, penology, and the design of daily life are currently subject to attack because they are said to threaten personal freedom, and I can testify that the attack can be fairly violent.

The extent to which a person is free or responsible for his achievements is not an issue to be decided by rigorous proof, but I submit that what we call the behaviour of the human organism is no more free than its digestion, gestation, immunization, or any other physiological process. Because it involves the environment in many subtle ways it is much more complex, and its lawfulness is, therefore, much harder to demonstrate. But a scientific analysis moves in that direction, and we can already throw some light on traditional topics, such as free will or creativity, which is more helpful than traditional accounts, and I believe that further progress is imminent.

The issue is, of course, determinism. Slightly more than a hundred years ago in a famous paper Claude Bernard raised with respect to physiology the issue which now stands before us in the behavioural sciences. The almost insurmountable obstacle to the application of scientific method in biology was, he said, the belief in 'vital spontaneity'. His contemporary, Louis Pasteur, was responsible for a dramatic test of the theory of spontaneous generation, and I suggest that the spontaneous generation of behaviour in the guise of ideas and acts of will is now at the stage of the spontaneous generation of life in the form of maggots and micro-organisms a hundred years ago.

The practical problem in continuing the struggle for freedom and dignity is not to destroy controlling forces but to change them, to create a world in which people will achieve far more than they have ever achieved before in art, music, literature, science, technology, and above all the enjoyment of life. It could be a world in which people feel freer than they have ever felt before, because they will not be under aversive control. In building such a world, we shall need all the help a science of behaviour can give us. To misread the theme of the struggle for freedom and dignity and to relinquish all efforts to control would be a tragic mistake.

But it is a mistake that may very well be made. Our concern for the individual as a creative agent is not dalliance; it is clearly an obstacle rather than a diversion. For ancient fears are not easily allayed. A shift in emphasis from the individual to the environment, particularly to the social environment, is reminiscent of various forms of totalitarian statism. It is easy to turn from what may seem like an inevitable movement in that direction and to take one's chances with libertarianism. But much remains to be analysed in that position. For example, we may distinguish between liberty and license by holding to the right to do as we please provided we do not infringe similar rights in others, but in doing so we conceal or disguise the public sanctions represented by private rights. Rights and duties, like a moral or ethical sense, are examples of hypothetical internalized environmental sanctions.

In the long run, the aggrandizement of the individual jeopardizes the future of the species and the culture. In effect it infringes the so-called rights of billions of people still to be born, in whose interests only the weakest of sanctions are now maintained. We are beginning to realize the magnitude of the problem of bringing human behaviour under the control of a projected future, and we are already suffering from the fact that we have come very late to recognize that mankind will have a future only if it designs a *viable* way of life. I wish I could share the optimism of both Darwin and Herbert Spencer that the course of evolution is necessarily towards perfection. It appears, on the contrary, that that course must be corrected from time to time. But, of course, if the intelligent behaviour that corrects it is also a product of evolution, then perhaps they were right after all. But it could be a near thing.

Perhaps it is now clear what I mean by diversions and obstacles. The science I am discussing is the investigation of the relation between behaviour and the environment—on the one hand, the environment in which the species evolved and which is responsible for the facts investigated by the ethologists and, on the other hand, the environment in which the individual lives and in response to which at any moment he behaves. We have been diverted from, and blocked in, our inquiries into the relations between behaviour and those environments by an absorbing interest in the organism itself. We have been misled by the almost instinctive tendency to look inside any system to see how it works, a tendency doubly powerful in the case of behaviour because of the apparent inside information supplied by feelings and introspectively observed states. Our only recourse is to leave that subject to the physiologist, who has, or will have, the only appropriate instruments and methods. We have also been encouraged to move in a centripetal direction because the discovery of controlling forces in the environment has seemed to reduce the credit due us for our achievements and to suggest that the struggle for freedom has not been as fully successful as we had imagined. We are not yet ready to accept the fact that the task is to change, not people, but rather the world in which they live.

We shall be less reluctant to abandon these diversions and to attack these obstacles, as we come to understand the possibility of a different approach. The role of the environment in human affairs has not, of course, gone unnoticed. Historians and biographers have acknowledged influences on human conduct, and literature has made the same point again and again. The Enlightenment advanced the cause of the individual by improving the world in which he lived—the Encyclopedia of Diderot and D'Alembert was designed to further changes of that sort—and by the nineteenth century, the controlling force of the environment was clearly recognized. Bentham and Marx have been called behaviourists, although for them the environment determined behaviour only after first determining consciousness, and this was an unfortunate qualification because the assumption of a mediating state clouded the relation between the terminal events.

The role of the environment has become clearer in the present century. Its selective action in evolution has been examined by the ethologists, and a similar selective action during the life of the individual is the subject of the experimental analysis of behaviour. In the current laboratory, very complex environments are constructed and their effects on behaviour studied. I believe this work offers consoling reassurance to those who are reluctant to abandon traditional formulations. Unfortunately, it is not well known outside the field. Its practical uses are, however, beginning to attract attention. Techniques derived from the analysis have proved useful in other parts of biology—for example, physiology and psychopharmacology—and have already led to the improved design of cultural practices, in programmed instructional materials, contingency management in the classroom, behavioural modification in psychotherapy and penology, and many other fields.

Much remains to be done, and it will be done more rapidly when the role of the environment takes its proper place in competition with the apparent evidences of an inner life. As Diderot put it, nearly two hundred years ago, 'Unfortunately it is easier and shorter to consult oneself than it is to consult nature. Thus the reason is inclined to dwell within itself.' But the problems we face are not to be found in men and women but in the world in which they live, especially in those social environments we call cultures. It is an important and promising shift in emphasis because, unlike the remote fastness of the so-called human spirit, the environment is within reach and we are learning how to change it.

And so I return to the role that has been assigned to me as a kind of twentieth-century Calvin, calling upon you to forsake the primrose path of total individualism, of self-actualization, self-adoration, and self-love, and to turn instead to the construction of that heaven on earth which is, I believe, within reach of the methods of science. I wish to testify that, once you are used to it, the way is not so steep or thorny after all.

NOTES

1. ONIANS, R. D. (1951). *The origins of European thought*. Cambridge University Press.

2. SHERRINGTON, C. S. (1906). *Integrative action of the nervous system*. Yale University Press, New Haven, Conn., U.S.A.

3. SKINNER, B. F. (1972). *Beyond freedom and dignity*. Jonathan Cape, London.

Preparation of this paper has been supported by a Career Award from the National Institute of Mental Health (Grant K6-MH-21,775-01).

0 The rationality of scientific revolutions

K. R. POPPER
University of London

THE title of this series of Spencer lectures, *Progress and obstacles to progress in the sciences,* was chosen by the organizers of the series. The title seems to me to imply that progress in science is a good thing, and that an obstacle to progress is a bad thing; a position held by almost everybody, until quite recently. Perhaps I should make clear at once that I accept this position, although with some slight and fairly obvious reservations to which I shall briefly allude later. Of course, obstacles which are due to the inherent difficulty of the problems tackled are welcome challenges. (Indeed, many scientists were greatly disappointed when it turned out that the problem of tapping nuclear energy was comparatively trivial, involving no new revolutionary change of theory.) But stagnation in science would be a curse. Still, I agree with Professor Bodmer's suggestion that scientific advance is only a *mixed* blessing.[1] Let us face it: blessings *are* mixed, with some exceedingly rare exceptions.

My talk will be divided into two parts. The first part (sections I–VIII) is devoted to progress in science, and the second part (sections IX–XIV) to some of the social obstacles to progress.

Remembering Herbert Spencer, I shall discuss progress in science largely *from an evolutionary point of view*; more precisely, from the point of view of the theory of natural selection. Only the end of the first part (that is, section VIII), will be spent in discussing the progress of science *from a logical point of view*, and in proposing *two rational criteria* of progress in science, which will be needed in the second part of my talk.

In the second part I shall discuss a few obstacles to progress in science, more especially ideological obstacles; and I shall end (sections XI–XIV) by discussing the distinction between, on the one hand, *scientific revolutions* which are subject to rational criteria of progress and, on the other hand, *ideological revolutions* which are only rarely rationally defensible. It appeared to me that this distinction was sufficiently interesting to call my lecture 'The rationality of scientific revolutions'. The emphasis here must be, of course, on the word 'scientific'.

I

I now turn to progress in science. I will be looking at progress in science from a biological or evolutionary point of view. I am far from suggesting that this is the most important point of view for examining progress in science. But the biological approach offers a convenient way of introducing the two leading ideas of the first half of my talk. They are the ideas of *instruction* and of *selection*.

From a biological or evolutionary point of view, science, or progress in science, may be regarded as a means used by the human species to adapt itself to the environment: to invade new environmental niches, and even to invent new environmental niches.[2] This leads to the following problem.

We can distinguish between three levels of adaptation: genetic adaptation; adaptive behavioural learning; and scientific discovery, which is a special case of adaptive behavioural learning. My main problem in this part of my talk will be to enquire into the similarities and dissimilarities between the strategies of progress or adaptation on the *scientific* level and on those two other levels: the *genetic* level and the *behavioural* level. And I will compare the three levels of adaptation by investigating the role played on each level by *instruction* and by *selection*.

II

In order not to lead you blindfolded to the result of this comparison I will anticipate at once my main thesis. It is a thesis asserting the *fundamental similarity of the three levels*, as follows.

On all three levels—genetic adaptation, adaptive behaviour, and scientific discovery—the mechanism of adaptation is fundamentally the same.

This can be explained in some detail.

Adaptation starts from an inherited *structure* which is basic for all three levels: *the gene structure of the organism*. To it corresponds, on the behavioural level, *the innate repertoire* of the types of behaviour which are available to the organism; and on the scientific level, *the dominant scientific conjectures or theories*. These *structures* are always transmitted by *instruction*, on all three levels: by the replication of the coded genetic instruction on the genetic and the behavioural levels; and by social tradition and imitation on the behavioural and the scientific levels. On all three levels, the *instruction* comes from *within the structure*. If mutations or variations or errors occur, then these are new instructions, which also arise *from within the structure*, rather than *from without*, from the environment.

These inherited structures are exposed to certain pressures, or challenges, or problems: to selection pressures; to environmental challenges; to theoretical problems. In response, variations of the genetically or traditionally inherited *instructions* are produced,[3] by methods which are at least partly *random*. On the genetic level, these are mutations and recombinations[4] of the coded instruction; on the behavioural level, they are tentative variations and recombinations within the repertoire; on the scientific level, they are new and revolutionary tentative theories. On all three levels we get new tentative trial instructions; or, briefly, tentative trials.

It is important that these tentative trials are changes that originate *within* the individual structure in a more or less random fashion— on all three levels. The view that they are *not* due to instruction from without, from the environment, is supported (if only weakly) by the fact that very similar organisms may sometimes respond in very different ways to the same new environmental challenge.

The next stage is that of *selection* from the available mutations and variations: those of the new tentative trials which are badly adapted are eliminated. *This is the stage of the elimination of error.* Only the more or less well adapted trial instructions survive and are inherited in their turn. Thus we may speak of *adaptation by 'the method of trial and error'* or better, by 'the method of trial and the elimination of error'. The elimination of error, or of badly adapted trial instructions, is also called '*natural selection*': it is a kind of 'negative feedback'. It operates on all three levels.

It is to be noted that in general *no equilibrium state of adaptation* is reached by any one application of the method of trial and the elimination of error, or by natural selection. First, because no perfect or optimal trial solutions to the problem are likely to be offered; secondly—and this is more important—because the emergence of new structures, or of new instructions, involves a change in the environmental situation. New elements of the environment may become relevant; and in consequence, new pressures, new challenges, new problems may arise, as a result of the structural changes which have arisen from within the organism.

On the genetic level the change may be a mutation of a gene, with a consequent change of an enzyme. Now the network of enzymes forms the more intimate environment of the gene structure. Accordingly, there will be a change in this intimate environment; and with it, new relationships between the organism and the more remote environment may arise; and further, new selection pressures.

The same happens on the behavioural level; for the adoption of a new kind of behaviour can be equated in most cases with the adoption of a new ecological niche. As a consequence, new selection pressures will arise, and new genetic changes.

On the scientific level, the tentative adoption of a new conjecture or theory may solve one or two problems, but it invariably opens up many *new* problems; for a new revolutionary theory functions exactly like a new and powerful sense organ. If the progress is significant then the new problems will differ from the old problems: the new problems will be on a radically different level of depth. This happened, for example, in relativity; it happened in quantum mechanics; and it happens right now, most dramatically, in molecular biology. In each of these cases, new horizons of unexpected problems were opened up by the new theory.

This, I suggest, is the way in which science progresses. And our progress can best be gauged by comparing our old problems with our new ones. If the progress that has been made is great, then the new problems will be of a character undreamt of before. There will be deeper problems; and besides, there will be more of them. The further we progress in knowledge, the more clearly we can discern the vastness of our ignorance.[5]

I will now sum up my thesis.

On all the three levels which I am considering, the genetic, the behavioural, and the scientific levels, we are operating with inherited structures which are passed on by instruction; either through the genetic code or through tradition. On all the three levels, new structures and new instructions arise by trial changes from *within the structure*: by tentative trials which are subject to natural selection or the elimination of error.

III

So far I have stressed the *similarities* in the working of the adaptive mechanism on the three levels. This raises an obvious problem: what about the *differences*?

The main difference between the genetic and the behavioural level is this. Mutations on the genetic level are not only random but completely 'blind', in two senses.[6] First, they are in no way goal directed. Secondly, the survival of a mutation cannot influence the further mutations, not even the frequencies or probabilities of their occurrence; though admittedly, the *survival* of a mutation may sometimes determine what kind of mutations may possibly *survive* in future cases. On the behavioural level, trials are also more or less random, but they are no longer completely 'blind' in either of the two senses mentioned. First, they are goal directed; and secondly, animals may learn from the outcome of a trial: they may learn to avoid the type of trial behaviour which has led to a failure. (They may even avoid it in cases in which it could have succeeded.) Similarly, they may also learn from success; and successful behaviour may be repeated, even in cases in which it is not adequate. However, a certain degree of 'blindness' is inherent in all trials.[7]

Behavioural adaptation is usually an intensely active process: the animal—especially the young animal at play—and even the plant, are actively investigating the environment.[8]

This activity, which is largely genetically programmed, seems to me to mark an important difference between the genetic level and the behavioural level. I may here refer to the experience which the *Gestalt* psychologists call 'insight'; an experience that accompanies many behavioural discoveries.[9] However, it must not be overlooked that even a discovery accompanied by 'insight' may be *mistaken*: every trial, even one with 'insight', is of the nature of a conjecture or a hypothesis. Köhler's apes, it will be remembered, sometimes hit with 'insight' on what turns out to be a mistaken attempt to solve their problem; and even great mathematicians are sometimes misled by intuition. Thus animals and men have to try out their hypotheses; they have to use the method of trial and of error elimination.

On the other hand I agree with Köhler and Thorpe[10] that the trials of problem-solving animals are in general not completely blind. Only in extreme cases, when the problem which confronts the animal does not yield to the making of hypotheses, will the animal resort to more or less blind and random attempts in order to get out of a disconcerting situation. Yet even in these attempts, goal-directedness is usually discernible, in sharp contrast to the blind randomness of genetic mutations and recombinations.

Another difference between genetic change and adaptive behavioural change is that the former *always* establishes a rigid and almost invariable genetic structure. The latter, admittedly, leads *sometimes* also to a fairly rigid behaviour pattern which is dogmatically adhered to; radically so in the case of 'imprinting' (Konrad Lorenz); but in other cases it leads to a flexible pattern which allows for differentiation or modification; for example, it may lead to exploratory behaviour, or to what Pavlov called the 'freedom reflex'.[11]

On the scientific level, discoveries are revolutionary and creative. Indeed, a certain creativity may be attributed to all levels, even to the genetic level: new trials, leading to new environments and thus to new selection pressures, create new and revolutionary results on all levels, even though there are strong conservative tendencies built into the various mechanisms of instruction.

Genetic adaptation can of course operate only within the time span of a few generations—at the very least, say, one or two generations. In organisms which replicate very quickly this may be a short time span; and there may be simply no room for behavioural adaptation. More slowly reproducing organisms are compelled to invent behavioural adaptation in order to adjust themselves to quick environmental changes. They thus need a behavioural repertoire, with types of behaviour of greater or lesser latitude or range. The repertoire, and the latitude of the available types of behaviour, can be assumed to be genetically programmed; and since, as indicated, a new type of behaviour may be said to involve the choice of a new environmental niche, new types of behaviour may indeed be genetically creative, for they may in their turn determine new selection pressures and thereby indirectly decide upon the future evolution of the genetic structure.[12]

On the level of scientific discovery two new aspects emerge. The most important one is that scientific theories can be formulated linguistically, and that they can even be published. Thus they become objects outside ourselves: objects open to investigation. As a consequence, they are now open to *criticism*. Thus we can get rid of a badly fitting theory before the adoption of the theory makes us unfit to survive: by criticizing our theories we can let our theories die in our stead. This is of course immensely important.

The other aspect is also connected with language. It is one of the novelties of human language that it encourages story telling, and thus *creative imagination*. Scientific discovery is akin to explanatory story telling, to myth making and to poetic imagination. The growth of imagination enhances of course the need for some control, such as, in science, inter-personal criticism—the friendly hostile co-operation of scientists which is partly based on competition and partly on the common aim to get nearer to the truth. This, and the role played by instruction and tradition, seems to me to exhaust the main sociological elements inherently involved in the progress of science; though more could be said of course about the social obstacles to progress, or the social dangers inherent in progress.

IV

I have suggested that progress in science, or scientific discovery, depends on *instruction* and *selection*: on a conservative or traditional or historical element, and on a revolutionary use of trial and the elimination of error by criticism, which includes severe empirical examinations or tests; that is, attempts to probe into the possible weaknesses of theories, attempts to refute them.

Of course, the individual scientist may wish to establish his theory rather than to refute it. But from the point of view of progress in science, this wish can easily mislead him. Moreover, if he does not himself examine his favourite theory critically, others will do so for him. The only results which will be regarded by them as supporting the theory will be the failures of interesting attempts to refute it; failures to find counter-examples where such counter-examples would be most expected, in the light of the best of the competing theories. Thus it need not create a great obstacle to science if the individual scientist is biased in favour of a pet theory. Yet I think that Claude Bernard was very wise when he wrote: 'Those who have an excessive faith in their ideas are not well fitted to make discoveries.'[13]

All this is part of the critical approach to science, as opposed to the inductivist approach; or of the Darwinian or eliminationist or selectionist approach as opposed to the Lamarckian approach which operates with the idea of *instruction from without*, or from the environment, while the critical or selectionist approach only allows *instruction from within*—from within the structure itself.

In fact, I contend that *there is no such thing as instruction from without the structure*, or the passive reception of a flow of information which impresses itself on our sense organs. All observations are theory impregnated: there is no pure, disinterested, theory-free observation. (To see this, we may try, using a little imagination, to compare human observation with that of an ant or a spider.)

Francis Bacon was rightly worried about the fact that our theories may prejudice our observations. This led him to advise scientists that they should avoid prejudice by purifying their minds of all theories. Similar recipes are still given.[14] But to attain objectivity we cannot rely on the empty mind: objectivity rests on criticism, on critical discussion, and on the critical examination of experiments.[15] And we must recognize, particularly, that our very sense organs incorporate what amount to prejudices. I have stressed before (in section II) that theories are like sense organs. Now I wish to stress that our sense organs are like theories. They *incorporate* adaptive theories (as has been shown in the case of rabbits and cats). And these theories are the result of natural selection.

V

However, not even Darwin or Wallace, not to mention Spencer, saw that there is no instruction from without. They did not operate with purely selectionist arguments. In fact, they frequently argued on Lamarckian lines.[16] In this they seem to have been mistaken. Yet it may be worthwhile to speculate about possible limits to Darwinism; for we should always be on the lookout for possible alternatives to any dominant theory.

I think that two points might be made here. The first is that the argument against the genetic inheritance of acquired characteristics (such as mutilations) depends upon the existence of a genetic mechanism in which there is a fairly sharp distinction between the gene structure and the remaining part of the organism: the soma. But this genetic mechanism must itself be a late product of evolution, and it was undoubtedly preceded by various other mechanisms of a less sophisticated kind. Moreover, certain very special kinds of mutilations *are* inherited; more particularly, mutilations of the gene structure by radiation. Thus if we assume that the primeval organism was a naked gene then we can even say that every non-lethal mutilation to this organism would be inherited. What we cannot say is that this fact contributes in any way to an explanation of genetic adaptation, or of genetic learning, except indirectly, via natural selection.

The second point is this. We may consider the very tentative conjecture that, as a somatic response to certain environmental pressures, some chemical mutagen is produced, increasing what is called the spontaneous mutation rate. This would be a kind of semi-Lamarckian effect, even though *adaptation* would still proceed only by the elimination of mutations; that is, by natural selection. Of course, there may not be much in this conjecture, as it seems that the spontaneous mutation rate suffices for adaptive evolution.[17]

These two points are made here merely as a warning against too dogmatic an adherence to Darwinism. Of course, I do conjecture that Darwinism is right, even on the level of scientific discovery; and that it is right even beyond this level: that it is right even on the level of artistic creation. We do not discover new facts or new effects by copying them, or by inferring them inductively from observation; or by any other method of instruction by the environment. We use, rather, the method of trial and the elimination of error. As Ernst Gombrich says, 'making comes before matching':[18] the active production of a new trial structure comes before its exposure to eliminating tests.

VI

I suggest therefore that we conceive the way science progresses somewhat on the lines of Niels Jerne's and Sir Macfarlane Burnet's theories of antibody formation.[19] Earlier theories of antibody formation assumed that the antigen works as a negative template for the formation of the antibody. This would mean that there is *instruction from without*, from the invading antibody. The fundamental idea of Jerne was that the instruction or information which enables the antibody to recognize the antigen is, literally, inborn: that it is part of the gene structure, though possibly subject to a repertoire of mutational variations. It is conveyed by the genetic code, by the chromosomes of the specialized cells which produce the antibodies; and the immune reaction is a result of growth-stimulation given to these cells by the antibody–antigen complex. Thus these cells are *selected* with the help of the invading environment (that is, with the help of the antigen), rather than instructed. (The analogy with the selection—and the modification—of scientific theories is clearly seen by Jerne, who in this connection refers to Kierkegaard, and to Socrates in the *Meno*.)

With this remark I conclude my discussion of the biological aspects of progress in science.

VII

Undismayed by Herbert Spencer's cosmological theories of evolution, I will now try to outline the cosmological significance of the contrast between *instruction from within the structure*, and *selection from without, by the elimination of trials*.

To this end we may note first the presence, in the cell, of the gene structure, the coded instruction, and of various chemical substructures;[20] the latter in random Brownian motion. The process of instruction by which the gene replicates proceeds as follows. The various substructures are carried (by Brownian motion) to the gene, in random fashion, and those which do not fit fail to attach themselves to the DNA structure; while those which fit, *do* attach themselves (with the help of enzymes). By this process of trial and selection,[21] a kind of photographic negative or complement of the genetic instruction is formed. Later, this complement separates from the original instruction; and by an analogous process, it forms again its negative. This negative of the negative becomes an identical copy of the original positive instruction.[22]

The selective process underlying replication is a fast-working mechanism. It is essentially the same mechanism that operates in most instances of chemical synthesis, and also, especially, in processes like crystallization. Yet although the underlying mechanism is selective, and operates by random trials and by the elimination of error, it functions as a part of what is clearly a process of instruction rather than of selection. Admittedly, owing to the random character of the motions involved, the matching processes will be brought about each time in a slightly different manner. In spite of this, the results are precise and conservative: the results are essentially determined by the original structure.

If we now look for similar processes on a cosmic scale, a strange picture of the world emerges which opens up many problems. It is a dualistic world: a world of structures in chaotically distributed motion. The small structures (such as the so-called elementary particles) build up larger structures; and this is brought about mainly by chaotic or random motion of the small structures, under special conditions of pressure and temperature. The larger structures may be atoms, molecules, crystals, organisms, stars, solar systems, galaxies, and galactic clusters. Many of these structures appear to have a seeding effect, like drops of water in a cloud, or crystals in a solution; that is to say, they can grow and multiply by instruction; and they may persist, or disappear by selection. Some of them, such as the aperiodic DNA crystals[23] which constitute the gene structure of organisms and, with it, their building instructions, are almost infinitely rare and, we may perhaps say, very precious.

I find this dualism fascinating: I mean the strange dualistic picture of a physical world consisting of comparatively stable structures— or rather structural processes—on all micro and macro levels; and of substructures on all levels, in apparently chaotically or randomly distributed motion: a random motion that provides part of the mechanism by which these structures and substructures are sustained, and by which they may seed, by way of instruction; and grow and multiply, by way of selection and instruction. This fascinating dualistic picture is compatible with, yet totally different from, the well-known dualistic picture of the world as indeterministic in the small, owing to quantum-mechanical indeterminism, and deterministic in the large, owing to macro-physical determinism. In fact, it looks as if the existence of structures which do the instructing, and which introduce something like stability into the world, depends very largely upon quantum effects.[24] This seems to hold for structures on the atomic, molecular, crystal, organic, and even on the stellar levels (for the stability of the stars depends upon nuclear reactions), while for the supporting random movements we can appeal to classical Brownian motion and to the classical hypothesis of molecular chaos. Thus in this dualist picture of order supported by disorder, or of structure supported by randomness, the role played by quantum effects and by classical effects appears to be almost the opposite from that in the more traditional pictures.

VIII

So far I have considered progress in science mainly from a biological point of view; however, it seems to me that the following two logical points are crucial.

First, in order that a new theory should constitute a discovery or a step forward it should conflict with its predecessor; that is to say, it should lead to at least some conflicting results. But this means, from a logical point of view, that it should contradict[25] its predecessor: it should overthrow it.

In this sense, progress in science—or at least striking progress—is always revolutionary.

My second point is that progress in science, although revolutionary rather than merely cumulative,[26] is in a certain sense always conservative: a new theory, however revolutionary, must always be able to explain fully the success of its predecessor. In all those cases in which its predecessor was successful, it must yield results at least as good as those of its predecessor and, if possible, better results. Thus in these cases the predecessor theory must appear as a good approximation to the new theory; while there should be, preferably, other cases where the new theory yields different and better results than the old theory.[27]

The important point about the two logical criteria which I have stated is that they allow us to decide of any new theory, even before it has been tested, whether it will be better than the old one, provided it stands up to tests. But this means that, in the field of science, we have something like a criterion for judging the quality of a theory as compared with its predecessor, and therefore a criterion of progress. And so it means that progress in science can be assessed rationally.[28] This possibility explains why, in science, only progressive theories are regarded as interesting; and it thereby explains why, as a matter of historical fact, the history of science is, by and large, a history of progress. (Science seems to be the only field of human endeavour of which this can be said.)

As I have suggested before, scientific progress is revolutionary. Indeed, its motto could be that of Karl Marx: 'Revolution in permanence.' However, scientific revolutions are rational in the sense that, in principle, it is rationally decidable whether or not a new theory is better than its predecessor. Of course, this does not mean that we cannot blunder. There are many ways in which we can make mistakes.

An example of a most interesting mistake is reported by Dirac.[29] Schrödinger found, but did not publish, a relativistic equation of the electron, later called the Klein–Gordon equation, before he found and published the famous non-relativistic equation which is now called by his name. He did not publish the relativistic equation because it did not seem to agree with the experimental results as interpreted by the preceding theory. However, the discrepancy was due to a faulty interpretation of empirical results, and not to a fault in the relativistic equation. Had Schrödinger published it, the problem of the equivalence between his wave mechanics and the matrix mechanics of Heisenberg and Born might not have arisen; and the history of modern physics might have been very different.

It should be obvious that the objectivity and the rationality of progress in science is not due to the personal objectivity and rationality of the scientist.[30] Great science and great scientists, like great poets, are often inspired by non-rational intuitions. So are great mathematicians. As Poincaré and Hadamard have pointed out,[31] a mathematical proof may be discovered by unconscious trials, guided by an inspiration of a decidedly aesthetic character, rather than by rational thought. This is true, and important. But obviously, it does not make the result, the mathematical proof, irrational. In any case, a proposed proof must be able to stand up to critical discussion: to its examination by competing mathematicians. And this may well induce the mathematical inventor to check, rationally, the results which he reached unconsciously or intuitively. Similarly, Kepler's beautiful Pythagorean dreams of the harmony of the world system did not invalidate the objectivity, the testability, the rationality of his three laws; nor the rationality of the problem which these laws posed for an explanatory theory.

With this, I conclude my two logical remarks on the progress of science; and I now move on to the second part of my lecture, and with it to remarks which may be described as partly sociological, and which bear on *obstacles* to progress in science.

IX

I think that the main obstacles to progress in science are of a social nature, and that they may be divided into two groups: economic obstacles and ideological obstacles.

On the economic side poverty may, trivially, be an obstacle (although great theoretical and experimental discoveries have been made in spite of poverty). In recent years, however, it has become fairly clear that affluence may also be an obstacle: too many dollars may chase too few ideas. Admittedly, even under such adverse circumstances progress *can* be achieved. But the spirit of science is in danger. Big Science may destroy great science, and the publication explosion may kill ideas: ideas, which are only too rare, may become submerged in the flood. The danger is very real, and it is hardly necessary to enlarge upon it, but I may perhaps quote Eugene Wigner, one of the early heroes of quantum mechanics, who sadly remarks:[32] 'The spirit of science has changed.'

This is indeed a sad chapter. But since it is all too obvious I shall not say more about the economic obstacles to progress in science; instead, I will turn to discuss some of the ideological obstacles.

X

The most widely recognized of the ideological obstacles is ideological or religious intolerance, usually combined with dogmatism and lack of imagination. Historical examples are so well known that I need not dwell upon them. Yet it should be noted that even suppression may lead to progress. The martyrdom of Giordano Bruno and the trial of Galileo may have done more in the end for the progress of science than the Inquisition could do against it.

The strange case of Aristarchus and the original heliocentric theory opens perhaps a different problem. Because of his heliocentric theory Aristarchus was accused of impiety by Cleanthes, a Stoic. But this hardly explains the obliteration of the theory. Nor can it be said that the theory was too bold. We know that Aristarchus's theory was supported, a century after it was first expounded, by at least one highly respected astronomer (Seleucus).[33] And yet, for some obscure reason, only a few brief reports of the theory have survived. Here is a glaring case of the only too frequent failure to keep alternative ideas alive.

Whatever the details of the explanation, the failure was probably due to dogmatism and intolerance. But new ideas should be regarded as precious, and should be carefully nursed; especially if they seem to be a bit wild. I do not suggest that we should be eager to accept new ideas *just* for the sake of their newness. But we should be anxious not to suppress a new idea even if it does not appear to us to be very good.

There are many examples of neglected ideas, such as the idea of evolution before Darwin, or Mendel's theory. A great deal can be learned about obstacles to progress from the history of these neglected ideas. An interesting case is that of the Viennese physicist Arthur Haas who in 1910 partly anticipated Niels Bohr. Haas published a theory of the hydrogen spectrum based on a quantization of J. J. Thomson's atom model: Rutherford's model did not yet exist. Haas appears to have been the first to introduce Planck's quantum of action into atomic theory with a view to deriving the spectral constants. In spite of his use of Thomson's atom model, Haas almost succeeded in his derivation; and as Max Jammer explains in detail, it seems quite possible that the theory of Haas (which was taken seriously by Sommerfeld) indirectly influenced Niels Bohr.[34] In Vienna, however, the theory was rejected out of hand; it was ridiculed, and decried as a silly joke by Ernst Lecher (whose early experiments had impressed Heinrich Hertz[35]), one of the professors of physics at the University of Vienna, whose somewhat pedestrian and not very inspiring lectures I attended some eight or nine years later.

A far more surprising case, also described by Jammer,[36] is the rejection in 1913 of Einstein's photon theory, first published in 1905, for which he was to receive the Nobel prize in 1921. This rejection of the photon theory formed a passage within a petition recommending Einstein for membership of the Prussian Academy of Science. The document, which was signed by Max Planck, Walther Nernst, and two other famous physicists, was most laudatory, and asked that a slip of Einstein's (such as they obviously believed his photon theory to be) should not be held against him. This confident manner of rejecting a theory which, in the same year, passed a severe experimental test undertaken by Millikan, has no doubt a humorous side; yet it should be regarded as a glorious incident in the history of science, showing that even a somewhat dogmatic rejection by the greatest living experts can go hand in hand with a most liberal-minded appreciation: these men did not dream of suppressing what they believed was mistaken. Indeed, the wording of the apology for Einstein's slip is most interesting and enlightening. The relevant passage of the petition says of Einstein: 'That he may sometimes have gone too far in his speculations, as for example in his hypothesis of light quanta, should not weigh too heavily against him. For nobody can introduce, even into the most exact of the natural sciences, ideas which are really new, without sometimes taking a risk.'[37] This is well said; but it is an understatement. One has always to take the risk of being mistaken, and also the less important risk of being misunderstood or misjudged.

However, this example shows, drastically, that even great scientists sometimes fail to reach that self-critical attitude which would prevent them from feeling very sure of themselves while gravely misjudging things.

Yet a limited amount of dogmatism is necessary for progress:

without a serious struggle for survival in which the old theories are tenaciously defended, none of the competing theories can show their mettle; that is, their explanatory power and their truth content. Intolerant dogmatism, however, is one of the main obstacles to science. Indeed, we should not only keep alternative theories alive by discussing them, but we should systematically look for new alternatives; and we should be worried whenever there are no alternatives— whenever a dominant theory becomes too exclusive. The danger to progress in science is much increased if the theory in question obtains something like a monopoly.

XI

But there is even a greater danger: a theory, even a scientific theory, may become an intellectual fashion, a substitute for religion, an entrenched ideology. And with this I come to the main point of this second part of my lecture—the part that deals with obstacles to progress in science: to the distinction between scientific revolutions and ideological revolutions.

For in addition to the always important problem of dogmatism and the closely connected problem of ideological intolerance, there is a different and, I think, a more interesting problem. I mean the problem which arises from certain links between science and ideology; links which do exist, but which have led some people to conflate science and ideology, and to muddle the distinction between scientific and ideological revolutions.

I think that this is quite a serious problem at a time when intellectuals, including scientists, are prone to fall for ideologies and intellectual fashions. This may well be due to the decline of religion, to the unsatisfied and unconscious religious needs of our fatherless society.[38] During my lifetime I have witnessed, quite apart from the various totalitarian movements, a considerable number of intellectually highbrow and avowedly non-religious movements with aspects whose religious character is unmistakable once your eyes are open to it.[39] The best of these many movements was that which was inspired by the father figure of Einstein. It was the best, because of Einstein's always modest and highly self-critical attitude and his humanity and tolerance. Nevertheless, I shall later have a few words to say about what seem to me the less satisfactory aspects of the Einsteinian ideological revolution.

I am not an essentialist, and I shall not discuss here the essence or nature of 'ideologies'. I will merely state very generally and vaguely that I am going to use the term 'ideology' for *any non-scientific* theory or creed or view of the world which proves attractive, and which interests people, including scientists. (Thus there may be very helpful and also very destructive ideologies from, say, a humanitarian or a rationalist point of view.[40]) I need not say more about ideologies in order to justify the sharp distinction which I am going to make between science[41] and 'ideology', and further, between *scientific revolutions* and *ideological revolutions*. But I will elucidate this distinction with the help of a number of examples.

These examples will show, I hope, that it is important to distinguish between a scientific revolution in the sense of a rational overthrow of an established scientific theory by a new one and all processes of 'social entrenchment' or perhaps 'social acceptance' of ideologies, including even those ideologies which incorporate some scientific results.

XII

As my first example I choose the Copernican and Darwinian revolutions, because in these two cases a scientific revolution gave rise to an ideological revolution. Even if we neglect here the ideology of 'Social Darwinism',[41a] we can distinguish a scientific and an ideological component in both these revolutions.

The Copernican and Darwinian revolutions were *ideological* in so far as they both changed man's view of his place in the Universe. They clearly were *scientific* in so far as each of them overthrew a dominant scientific theory: a dominant astronomical theory and a dominant biological theory.

It appears that the ideological impact of the Copernican and also of the Darwinian theory was so great because each of them clashed with a religious dogma. This was highly significant for the intellectual history of our civilization, and it had repercussions on the history of science (for example, because it led to a tension between religion and science). And yet, the historical and sociological fact that the theories of both Copernicus and Darwin clashed with religion is completely irrelevant for the rational evaluation of the scientific theories proposed by them. Logically it has nothing whatsoever to do with the *scientific* revolution started by each of the two theories.

It is therefore important to distinguish between scientific and ideological revolutions particularly in those cases in which the ideological revolutions interact with revolutions in science.

The example, more especially, of the Copernican ideological revolution may show that even an ideological revolution might well be described as 'rational'. However, while we have a logical criterion of progress in science—and thus of rationality—we do not seem to have anything like general criteria of progress or of rationality outside science (although this should not be taken to mean that outside science there are no such things as standards of rationality). Even a highbrow intellectual ideology which bases itself on accepted scientific results may be irrational, as is shown by the many movements of modernism in art (and in science), and also of archaism in art; movements which in my opinion are intellectually insipid since they appeal to values which have nothing to do with art (or science). Indeed, many movements of this kind are just fashions which should not be taken too seriously.[42]

Proceeding with my task of elucidating the distinction between scientific and ideological revolutions, I will now give several examples of major scientific revolutions which did not lead to any ideological revolution.

The revolution of Faraday and Maxwell was, from a scientific point of view, just as great as that of Copernicus, and possibly greater: it dethroned Newton's central dogma—the dogma of central forces. Yet it did *not* lead to an ideological revolution, though it inspired a whole generation of physicists.

J. J. Thomson's discovery (and theory) of the electron was also a major revolution. To overthrow the age-old theory of the indivisibility of the atom constituted a scientific revolution easily comparable to Copernicus's achievement: when Thomson announced it, physicists thought he was pulling their legs. But it did not create an ideological revolution. And yet, it overthrew both of the two rival theories which for 2400 years had been fighting for dominance in the theory of matter—the theory of indivisible atoms, and that of the continuity of matter. To assess the revolutionary significance of this breakthrough it will be sufficient to remind you that it introduced structure as well as electricity into the atom, and thus into the constitution of matter. Also, the quantum mechanics of 1925 and 1926, of Heisenberg and of Born, of de Broglie, of Schrödinger and of Dirac, was essentially a quantization of the theory of the Thomson electron. And yet Thomson's scientific revolution did not lead to a new ideology.

Another striking example is Rutherford's overthrow in 1911 of the

model of the atom proposed by J. J. Thomson in 1903. Rutherford
had accepted Thomson's theory that the positive charge must be
distributed over the whole space occupied by the atom. This may be
seen from his reaction to the famous experiment of Geiger and
Marsden. They found that when they shot alpha particles at a very
thin sheet of gold foil, a few of the alpha particles—about one in
twenty thousand—were reflected by the foil rather than merely
deflected. Rutherford was incredulous. As he said later:[43] 'It was
quite the most incredible event that has ever happened to me in my
life. It was almost as incredible as if you fired a fifteen-inch shell at
a piece of tissue paper and it came back and hit you.' This remark of
Rutherford's shows the utterly revolutionary character of the discovery. Rutherford realized that the experiment refuted Thomson's
model of the atom, and he replaced it by his nuclear model of the
atom. This was the beginning of nuclear science. Rutherford's
model became widely known, even among non-physicists. But it did
not trigger off an ideological revolution.

One of the most fundamental scientific revolutions in the history
of the theory of matter has never even been recognized as such. I
mean the refutation of the electromagnetic theory of matter which
had become dominant after Thomson's discovery of the electron.
Quantum mechanics arose as part of this theory, and it was essentially
this theory whose 'completeness' was defended by Bohr against
Einstein in 1935, and again in 1949. Yet in 1934 Yukawa had outlined
a new quantum-theoretical approach to nuclear forces which resulted
in the overthrow of the electromagnetic theory of matter, after forty
years of unquestioned dominance.[44]

There are many other major scientific revolutions which failed to
trigger off any ideological revolution; for example, Mendel's revolution (which later saved Darwinism from extinction). Others are
X-rays, radio-activity, the discovery of isotopes, and the discovery of
superconductivity. To all these, there was no corresponding ideological revolution. Nor do I see as yet an ideological revolution
resulting from the breakthrough of Crick and Watson.

XIII

Of great interest is the case of the so-called Einsteinian revolution;
I mean Einstein's scientific revolution which among intellectuals
had an ideological influence comparable to that of the Copernican or
Darwinian revolutions.

Of Einstein's many revolutionary discoveries in physics, there are two which are relevant here.

The first is special relativity, which overthrows Newtonian kinematics, replacing Galileo invariance by Lorentz invariance.[45] Of course, this revolution satisfies our criteria of rationality: the old theories are explained as approximately valid for velocities which are small compared with the velocity of light.

As to the ideological revolution linked with this scientific revolution, one element of it is due to Minkowski. We may state this element in Minkowski's own words. 'The views of space and time I wish to lay before you', Minkowski wrote, '. . . are radical. Henceforth space by itself, and time by itself, are doomed to fade away into mere shadows, and only a kind of union of the two will preserve an independent reality.'[46] This is an intellectually thrilling statement. But it is clearly not science: it is ideology. It became part of the ideology of the Einsteinian revolution. But Einstein himself was never quite happy about it. Two years before his death he wrote to Cornelius Lanczos: 'One knows so much and comprehends so little. The four-dimensionality with the [Minkowski signature of] $+ + + -$ belongs to the latter category.'

A more suspect element of the ideological Einsteinian revolution is the fashion of operationalism or positivism—a fashion which Einstein later rejected, although he himself was responsible for it, owing to what he had written about the operational definition of simultaneity. Although, as Einstein later realized,[47] operationalism is, logically, an untenable doctrine, it has been very influential ever since, in physics, and especially in behaviourist psychology.

With respect to the Lorentz transformations, it does not seem to have become part of the ideology that they limit the validity of the transitivity of simultaneity: the principle of transitivity remains valid within each inertial system while it becomes invalid for the transition from one system to another. Nor has it become part of the ideology that general relativity, or more especially Einstein's cosmology, allows the introduction of a preferred cosmic time and consequently of preferred local spatio-temporal frames.[48]

General relativity was in my opinion one of the greatest scientific revolutions ever, because it clashed with the greatest and best tested theory ever—Newton's theory of gravity and of the solar system. It contains, as it should, Newton's theory as an approximation; yet it contradicts it in several points. It yields different results for elliptic orbits of appreciable eccentricity; and it entails the astonishing result that any physical particle (photons included) which approaches the centre of a gravitational field with a velocity exceeding six-tenths of the velocity of light is not accelerated by the gravitational field, as in Newton's theory, but decelerated: that is, not attracted by a heavy body, but repelled.[49]

This most surprising and exciting result has stood up to tests; but it does not seem to have become part of the ideology.

It is this overthrow and correction of Newton's theory which from a scientific (as opposed to an ideological) point of view is perhaps most significant in Einstein's general theory. This implies, of course, that Einstein's theory can be compared point by point with Newton's[50] and that it preserves Newton's theory as an approximation. Nevertheless, Einstein never believed that his theory was true. He shocked Cornelius Lanczos in 1922 by saying that his theory was merely a passing stage: he called it 'ephemeral'.[51] And he said to Leopold Infeld[52] that the left-hand side of his field equation[53] (the curvature tensor) was as solid as a rock, while the right-hand side (the momentum–energy tensor) was as weak as straw.

In the case of general relativity, an idea which had considerable ideological influence seems to have been that of a curved four-dimensional space. This idea certainly plays a role in both the scientific and the ideological revolution. But this makes it even more important to distinguish the scientific from the ideological revolution.

However, the ideological elements of the Einsteinian revolution influenced scientists, and thereby the history of science; and this influence was not all to the good.

First of all, the myth that Einstein had reached his result by an essential use of epistemological and especially operationalist methods had in my opinion a devastating effect upon science. (It is irrelevant whether you get your results—especially good results—by dreaming them, or by drinking black coffee, or even from a mistaken epistemology.[53a]) Secondly, it led to the belief that quantum mechanics, the second great revolutionary theory of the century, must outdo the Einsteinian revolution, especially with respect to its epistemological depth. It seems to me that this belief affected some of the great founders of quantum mechanics,[54] and also some of the great founders of molecular biology.[55] It led to the dominance of a subjectivist interpretation of quantum mechanics; an interpretation which I have been combating for almost forty years. I cannot here describe the situation; but while I am aware of the dazzling achievement of quantum mechanics (which must not blind us to the fact that it is seriously incomplete[56]) I suggest that the orthodox interpretation of quantum mechanics is not part of physics, but an ideology. In fact, it is part of a modernistic ideology; and it has become a scientific fashion which is a serious obstacle to the progress of science.

XIV

I hope that I have made clear the distinction between a scientific revolution and the ideological revolution which may sometimes be linked with it. The ideological revolution may serve rationality or it may undermine it. But it is often nothing but an intellectual fashion. Even if it is linked to a scientific revolution it may be of a highly irrational character; and it may consciously break with tradition.

But a scientific revolution, however radical, cannot really break with tradition, since it must preserve the success of its predecessors. This is why scientific revolutions are rational. By this I do not mean, of course, that the great scientists who make the revolution are, or ought to be, wholly rational beings. On the contrary: although I have been arguing here for the rationality of scientific revolutions, my guess is that should individual scientists ever become 'objective and rational' in the sense of 'impartial and detached', then we should indeed find the revolutionary progress of science barred by an impenetrable obstacle.

NOTES

1. See, in the present series of Herbert Spencer Lectures, the concluding remark of the contribution by Professor W. F. Bodmer. My own misgivings concerning scientific advance and stagnation arise mainly from the changed spirit of science, and from the unchecked growth of Big Science which endangers great science. (See section IX of this lecture.) Biology seems to have escaped this danger so far, but not, of course, the closely related dangers of large-scale application.

2. The formation of membrane proteins, of the first viruses, and of cells, may perhaps have been among the earliest inventions of new environmental niches; though it is possible that other environmental niches (perhaps networks of enzymes invented by otherwise naked genes) may have been invented even earlier.

3. It is an open problem whether one can speak in these terms ('in response') about the genetic level (compare my conjecture about responding mutagens in section V). Yet if there were no variations, there could not be adaptation or evolution; and so we can say that the occurrence of mutations is either partly controlled by a need for them, or functions as if it was.

4. When in this lecture I speak, for brevity's sake, of 'mutation'; the possibility of recombination is of course always tacitly included.

5. The realization of our ignorance has become pinpointed as a result, for example, of the astonishing revolution brought about by molecular biology.

6. For the use of the term 'blind' (especially in the second sense) see D. T. Campbell, Methodological suggestions from a comparative psychology of knowledge processes, *Inquiry* **2**, 152–82 (1959); Blind variation and selective retention in creative thought as in other knowledge processes, *Psychol. Rev.* **67**, 380–400 (1960); and Evolutionary epistemology, in *The philosophy of Karl Popper*, The library of living philosophers (ed. P. A. Schilpp), pp. 413–63, The Open Court Publishing Co., La Salle, Illinois (1974).

7. While the 'blindness' of trials is relative to what we have found out in the past, randomness is relative to a set of elements (forming the 'sample space'). On the genetic level these 'elements' are the four nucleotide bases; on the behavioural level they are the constituents of the organism's repertoire of behaviour. These constituents may assume different weights with respect to different needs or goals, and the weights may change through experience (lowering the degree of 'blindness').

8. On the importance of active participation, see R. Held and A. Hein, Movement-produced stimulation in the development of visually guided behaviour, *J. comp. Physiol. Psychol.* **56**, 872–6 (1963); cf. J. C. Eccles, *Facing reality*, pp. 66–7. The activity is, at least partly, one of producing hypotheses: see J. Krechevsky, 'Hypothesis' versus 'chance' in the pre-solution period in sensory discrimination-learning, *Univ. Calif. Publ. Psychol.* **6**, 27–44 (1932) (reprinted in *Animal problem solving* (ed. A. J. Riopelle), pp. 183–97, Penguin Books, Harmondsworth (1967))

9. I may perhaps mention here some of the differences between my views and the views of the *Gestalt* school. (Of course, I accept the fact of *Gestalt* perception; I am only dubious about what may be called *Gestalt* philosophy.)

I conjecture that the unity, or the articulation, of perception is more closely dependent on the motor control systems and the efferent neural systems of the brain than on afferent systems: that it is closely dependent on the behavioural repertoire of the organism. I conjecture that a spider or a mouse will never have insight (as had Köhler's ape) into the possible unity of the two sticks which can be joined together, because handling sticks of that size does not belong to their behavioural repertoire. All this may be interpreted as a kind of generalization of the James–Lange theory of emotions (1884; see William James, *The principles of psychology*, Vol. II, pp. 449 ff. (1890) Macmillan and Co., London), extending the theory from our emotions to our perceptions (especially to *Gestalt* perceptions) which thus would not be 'given' to us (as in *Gestalt* theory) but rather 'made' by us, by decoding (comparatively 'given') clues. The fact that the clues may mislead (optical illusions in man; dummy illusions in animals, etc.) can be explained by the biological need to impose our behavioural interpretations upon highly simplified clues. The conjecture that our decoding of what the senses tell us depends on our behavioural repertoire may explain part of the gulf that lies between animals and man; for through the evolution of the human language our repertoire has become unlimited.

10. See W. H. Thorpe, *Learning and instinct in animals*, pp. 99 ff. Methuen, London (1956); 1963 edn, pp. 100–47; W. Köhler, *The mentality of apes* (1925); Penguin Books edn, (1957), pp. 166 ff.

11. See I. P. Pavlov, *Conditioned reflexes*, esp. pp. 11–12, Oxford University Press (1927). In view of what he calls 'exploratory behaviour' and the closely related 'freedom behaviour'—both obviously genetically based—and of the significance of these for scientific activity, it seems to me that the behaviour of behaviourists who aim to supersede the value of freedom by what they call 'positive reinforcement' may be a symptom of an unconscious hostility to science.

Incidentally, what B. F. Skinner (cf. his *Beyond freedom and dignity* (1972) Cape, London) calls 'the literature of freedom' did not arise as a result of negative reinforcement, as he suggests. It arose, rather, with Aeschylus and Pindar, as a result of the victories of Marathon and Salamis.

12. Thus exploratory behaviour and problem solving create new conditions for the evolution of genetic systems; conditions which deeply affect the natural selection of these systems. One can say that once a certain latitude of behaviour has been attained—as it has been attained even by unicellular organisms (see especially the classic work of H. S. Jennings, *The behaviour of the lower organisms*, Columbia University Press, New York (1906))—the initiative of the organism in selecting its ecology or habitat takes the lead, and natural selection within the new habitat follows the lead. In this way, Darwinism can simulate Lamarckism, and even Bergson's 'creative evolution'. This has been recognized by strict Darwinists. For a brilliant presentation and survey of the history, see Sir Alister Hardy, *The living stream*, Collins, London (1965), especially lectures VI, VII, and VIII, where many references to earlier literature will be found, from James Hutton (who died in 1797) onwards (see pp. 178 ff.). See also Ernst Mayr, *Animal species and evolution*, The Belknap Press, Cambridge, Mass., and Oxford University Press, London (1963), pp. 604 ff. and 611; Erwin Schrödinger, *Mind and Matter*, Cambridge University Press (1958), ch. 2; F. W. Braestrup, The evolutionary significance of learning, in *Vidensk. Meddr dansk naturh. Foren.* **134**, 89–102 (1971) (with a bibliography); and also my first Herbert Spencer Lecture (1961) now in my *Objective knowledge*, Clarendon Press, Oxford (1972, 1973).

13. Quoted by Jacques Hadamard, *The psychology of invention in the mathematical field*, Princeton University Press (1945), and Dover edition (1954), p. 48.

14. Behavioural psychologists who study 'experimenter bias' have found that some albino rats perform decidedly better than others if the experimenter is led to believe (wrongly) that the former belong to a strain selected for high intelligence: see, The effect of experimenter bias on the performance of the albino rat, *Behav. Sci.* **8**, 183–9 (1963). The lesson drawn by the authors of this paper is that experiments should be made by 'research assistants who do not know what outcome is desired' (p. 188). Like Bacon, these authors pin their hopes on the empty mind, forgetting that the expectations of the director of research may communicate themselves, without explicit disclosure, to his research assistants, just as they seem to have communicated themselves from each research assistant to his rats.

15. Compare my *Logic of scientific discovery*, section 8, and my *Objective knowledge*.

16. It is interesting that Charles Darwin in his later years believed in the occasional inheritance even of mutilations. See his *The variation of animals and plants under domestication*, 2nd edn, Vol. i, pp. 466–70 (1875).

17. Specific mutagens (acting selectively, perhaps on some particular sequence of codons rather than on others) are not known, I understand. Yet their existence would hardly be surprising in this field of surprises; and they might explain mutational 'hot spots'. At any rate, there seems to be a real difficulty in concluding from the absence of known specific mutagens that specific mutagens do not exist. Thus it seems to me that the problem suggested in the text (the possibility of a reaction to certain strains by the production of mutagens) is still open.

18. Cf. Ernst Gombrich, *Art and illusion* (1960), and later editions. (See the Index under 'making and matching'.)

19. See Niels Kai Jerne, The natural selection theory of antibody formation; ten years later, in *Phage and the origin of molecular biology* (ed. J. Cairns *et. al.*), pp. 301–12 (1966); also The natural selection theory of antibody formation,

Proc. natn. Acad. Sci. **41**, 849–57 (1955); Immunological speculations, *A. Rev. Microbiol.* **14**, 341–58 (1960); The immune system, *Scient. Am.* **229**, 52–60. See also Sir Macfarlane Burnet, A modification of Jerne's theory of anti-body production, using the concept of clonal selection, *Aust. J. Sci.* **20**, 67–9 (1957); *The clonal selection theory of acquired immunity*, Cambridge University Press (1959).

20. What I call 'structures' and 'substructures' are called 'integrons' by Francois Jacob, *The logic of living systems: a history of heredity*, pp. 299–324. Allen Lane, London (1974).

21. Something might be said here about the close connection between 'the method of trial and of the elimination of error' and 'selection': all selection is error elimination; and what remains—after elimination—as 'selected' are merely those trials which have not been eliminated *so far*.

22. The main difference from a photographic reproduction process is that the DNA molecule is not two-dimensional but linear: a long string of four kinds of substructures ('bases'). These may be represented by dots coloured either red or green, or blue or yellow. The four basic colours are pairwise negatives (or complements) of each other. So the negative or complement of a string would consist of a string in which red is replaced by green, and blue by yellow; and vice versa. Here the colours represent the four letters (bases) which constitute the alphabet of the genetic code. Thus the complement of the original string contains a kind of translation of the original information into another yet closely related code; and the negative of this negative contains in turn the original information, stated in terms of the original (the genetic) code.

This situation is utilized in replication, when first one pair of complementary strings separates and when next two pairs are formed as each of the strings selectively attaches to itself a new complement. The result is the replication of the original structure, *by way of instruction*. A very similar method is utilized in the second of the two main functions of the gene (DNA): the control, by way of instruction, of the synthesis of proteins. Though the underlying mechanism of this second process is more complicated than that of replication, it is similar in principle.

23. The term 'aperiodic crystal' (sometimes also 'aperiodic solid') is Schrö-dinger's; see his *What is life?*, Cambridge University Press (1944); cf. *What is life? and Mind and matter*, Cambridge University Press, pp. 64 and 91 (1967).

24. That atomic and molecular structures have something to do with quantum theory is almost trivial, considering that the peculiarities of quantum mechanics (such as eigenstates and eigenvalues) were introduced into physics in order to explain the structural stability of atoms.

The idea that the structural 'wholeness' of biological systems has also something to do with quantum theory was first discussed, I suppose, in Schrödinger's small but great book *What is life?* (1944) which, it may be said, anticipated both the rise of molecular biology and of Max Delbrück's influence on its development. In this book Schrödinger adopts a consciously ambivalent attitude towards the problem whether or not biology will turn out to be reducible to physics. In Chapter 7, 'Is life based on the laws of physics', he says (about living matter) first that 'we must be prepared to find it working in a manner that cannot be reduced to the ordinary laws of physics' (*What is life? and Mind and matter*, p. 81). But a little later he says that 'the new principle' (that is to say, 'order from order') 'is not alien to physics': it is 'nothing else than the principle of quantum physics again' (in the form of Nernst's principle) (*What is life? and Mind and matter*, p. 88). My attitude is also an ambivalent one: on the one hand, I do not believe in complete reducibility; on the other hand, I think that *reduction must be attempted*;

for even though it is likely to be only partially successful, even a very partial success would be a very great success.

Thus my remarks in the text to which this note is appended (and which I have left substantially unchanged) were not meant as a statement of reductionism: all I wanted to say was that quantum theory seems to be involved in the phenomenon of 'structure from structure' or 'order from order'.

However, my remarks were not clear enough; for in the discussion after the lecture Professor Hans Motz challenged what he believed to be my reductionism by referring to one of the papers of Eugene Wigner ('The probability of the existence of a self-reproducing unit', ch. 15 of his *Symmetries and reflections: scientific essays*, pp. 200–8, M.I.T. Press (1970)). In this paper Wigner gives a kind of proof of the thesis that the probability is zero for a quantum theoretical system to contain a subsystem which reproduces itself. (Or, more precisely, the probability is zero for a system to change in such a manner that at one time it contains some subsystem and later a second subsystem which is a copy of the first.) I have been puzzled by this argument of Wigner's since its first publication in 1961; and in my reply to Motz I pointed out that Wigner's proof seemed to me refuted by the existence of Xerox machines (or by the growth of crystals) which must be regarded as quantum mechanical rather than 'biotonic' systems. (It may be claimed that a Xerox copy or a crystal does not reproduce itself with sufficient precision; yet the most puzzling thing about Wigner's paper is that he does not refer to degrees of precision, and that absolute exactness or 'the apparently virtually absolute reliability' as he puts it on p. 208—which is not required—is, it seems, excluded at once by Pauli's principle.) I do not think that either the reducibility of biology to physics or else its irreducibility can be proved; at any rate not at present.

25. Thus Einstein's theory *contradicts* Newton's theory (although it contains Newton's theory as an approximation): in contradistinction to Newton's theory, Einstein's theory shows for example that in strong gravitational fields there cannot be a Keplerian elliptic orbit with appreciable eccentricity but without corresponding precession of the perihelion (as observed of Mercury).

26. Even the collecting of butterflies is *theory*-impregnated ('butterfly' is a *theoretical* term, as is 'water': it involves a set of expectations). The recent accumulation of evidence concerning elementary particles can be interpreted as an accumulation of falsifications of the early electromagnetic theory of matter.

27. An even more radical demand may be made; for we may demand that if the apparent laws of nature should change, then the new theory, invented to explain the new laws, should be able to explain the state of affairs both before and after the change, and also the change itself, from universal laws and (changing) initial conditions (cf. my *Logic of scientific discovery*, section 79, esp. p. 253).

By stating these logical criteria for progress, I am implicitly rejecting the fashionable (anti-rationalistic) suggestion that two different theories such as Newton's and Einstein's are incommensurable. It may be true that two scientists with a verificationist attitude towards their favoured theories (Newtonian and Einsteinian physics, say) may fail to understand each other. But if their attitude is critical (as was Newton's and Einstein's) they will understand both theories, and see how they are related. See, for this problem, the excellent discussion of the comparability of Newton's and Einstein's theories by Troels Eggers Hansen in his paper, Confrontation and objectivity, *Danish Yb. Phil.* 7, 13–72 (1972).

28. The logical demands discussed here (cf. ch. 10 of my *Conjectures and refutations* and ch. 5 of *Objective knowledge*), although they seem to me of fundamental importance, do not, of course, exhaust what can be said about the rational method of science. For example, in my *Postscript* (which has been in galley proofs since 1957, but which I hope will still be published one day) I have

developed a theory of what I call metaphysical research programmes. This theory, it might be mentioned, in no way clashes with the theory of testing and of the revolutionary advance of science which I have outlined here. An example which I gave there of a metaphysical research programme is the use of the propensity theory of probability, which seems to have a wide range of applications.

What I say in the text should not be taken to mean that rationality depends on having a criterion of rationality. Compare my criticism of 'criterion philosophies in Addendum I, Facts, standards, and truth, to Vol. ii of my *Open society*.

29. The story is reported by Paul A. M. Dirac, The evolution of the physicist's picture of nature, *Scient. Am.* **208**, No. 5, 45–53 (1963); see esp. p. 47.

30. Cf. my criticism of the so-called 'sociology of knowledge' in ch. 23 of my *Open society*, and pp. 155 f. of my *Poverty of historicism*.

31. Cf. Jacques Hadamard, *The psychology of invention in the mathematical field* (see note 13 above).

32. A conversation with Eugene Wigner, *Science* **181**, 527–33 (1973); see p. 533.

33. For Aristarchus and Seleucus see Sir Thomas Heath, *Aristarchus of Samos*, Clarendon Press, Oxford (1966).

34. See Max Jammer, *The conceptual development of quantum mechanics*, pp. 40–2, McGraw-Hill, New York (1966).

35. See Heinrich Hertz, *Electric waves*, Macmillan and Co., London (1894); Dover edn, New York (1962), pp. 12, 187 f., 273.

36. See Jammer, *op. cit.*, pp. 43 f., and Théo Kahan, Un document historique de l'académie des sciences de Berlin sur l'activité scientifique d'Albert Einstein (1913), *Archs. int. Hist. Sci.* **15**, 337–42 (1962); see esp. p. 340.

37. Compare Jammer's slightly different translation, *loc. cit.*

38. Our Western societies do not, by their structure, satisfy the need for a father-figure. I discussed the problems that arise from this fact briefly in my (unpublished) William James Lectures in Harvard (1950). My late friend, the psychoanalyst Paul Federn, showed me shortly afterwards an earlier paper of his devoted to this problem.

39. An obvious example is the role of prophet played, in various movements, by Sigmund Freud, Arnold Schönberg, Karl Kraus, Ludwig Wittgenstein, and Herbert Marcuse.

40. There are many kinds of 'ideologies' in the wide and (deliberately) vague sense of the term I used in the text, and therefore many aspects to the distinction between science and ideology. Two may be mentioned here. One is that scientific theories can be distinguished or 'demarcated' (see note 41) from non-scientific theories which, nevertheless, may strongly influence scientists, and even inspire their work. (This influence, of course, may be good or bad or mixed.) A very different aspect is that of entrenchment: a scientific theory may function as an ideology if it becomes socially entrenched. This is why, when speaking of the distinction between scientific revolutions and ideological revolutions, I include among ideological revolutions changes in non-scientific ideas which may inspire the work of scientists, and also changes in the social entrenchment of what may otherwise be a scientific theory. (I owe the formulation of the points in this note to Jeremy Shearmur who has also contributed to other points dealt with in this lecture.)

41. In order not to repeat myself too often, I did not mention in this lecture my suggestion for a criterion of the empirical character of a theory (falsifiability or refutability as the criterion of demarcation between empirical theories and non-empirical theories). Since in English 'science' means 'empirical science', and since the matter is sufficiently fully discussed in my books, I have written things like the following (for example, in *Conjectures and refutations*, p. 39): '. . . in order to

be ranked as scientific, [statements] must be capable of conflicting with possible, or conceivable, observations.' Some people seized upon this like a shot (as early as 1932, I think). 'What about your own gospel?' is the typical move. (I found this objection again in a book published in 1973.) My answer to the objection, however, was published in 1934 (see my *Logic of scientific discovery*, ch. 2, section 10 and elsewhere). I may restate my answer: my gospel is not 'scientific', that is, it does not belong to empirical science, but it is, rather, a (normative) *proposal*. My gospel (and also my answer) is, incidentally, criticizable, though not just by observation; and it has been criticized.

41a. For a criticism of Social Darwinism see my *Open Society*, ch. 10, note 71.

42. Further to my use of the vague term 'ideology' (which includes all kinds of theories, beliefs, and attitudes, including some that may influence scientists) it should be clear that I intend to cover by this term not only historicist fashions like 'modernism', but also serious, and rationally discussable, metaphysical and ethical ideas. I may perhaps refer to Jim Erikson, a former student of mine in Christchurch, New Zealand, who once said in a discussion: 'We do not suggest that science invented intellectual honesty, but we do suggest that intellectual honesty invented science.' A very similar idea is to be found in ch. ix (The kingdom and the darkness) of Jacques Monod's book *Chance and necessity*, Knopf, New York (1971). See also my *Open society*, vol. ii, ch. 24 (The revolt against reason). We might say, of course, that an ideology which has learned from the critical approach of the sciences is likely to be more rational than one which clashes with science.

43. Lord Rutherford, The development of the theory of atomic structure, in J. Needham and W. Pagel (eds), *Background of modern science*, pp. 61–74, Cambridge University Press (1938); the quotation is from p. 68.

44. See my Quantum mechanics without 'the observer', in *Quantum theory and reality* (ed. Mario Bunge, esp. pp. 8–9, Springer-Verlag, New York (1967). (It will form a chapter in my forthcoming volume *Philosophy and physics*.)

The fundamental idea (that the inertial mass of the electron is in part explicable as the inertia of the moving electromagnetic field) which led to the electromagnetic theory of matter is due to J. J. Thomson, On the electric and magnetic effects produced by the motion of electrified bodies, *Phil. Mag.* (5th Ser.) **11**, 229–49 (1881), and to O. Heaviside, On the electromagnetic effects due to the motion of electrification through a dialectric, *Phil. Mag.* (5th Ser.) **27**, 324–39 (1889). It was developed by W. Kaufmann (Die magnetische und elektrische Ablenkbarkeit der Bequerelstrahlen und die scheinbare Masse der Elektronen, *Gött. Nachr.* 143–55 (1901), Ueber die elektromagnetische Masse des Elektrons, 291–6 (1902), Ueber die 'Elektromagnetische Masse' der Elektronen, 90–103 (1903)) and M. Abraham (Dynamik des Elektrons, *Gött. Nachr.*, 20–41 (1902), Prinzipien der Dynamik des Elektrons, *Annln Phys.* (4th Ser.), **10**, 105–79 (1903)) into the thesis that the mass of the electron is a purely electromagnetic effect. (See W. Kaufmann, Die elektromagnetische Masse des Elektrons, *Phys. Z.* **4**, 54–7 (1902–3) and M. Abraham, Prinzipien der Dynamik des Elektrons, *Phys. Z.* **4**, pp. 57–63 (1902–3) and M. Abraham, *Theorie der Elektrizität*, Vol. ii, pp. 136–249, Leipzig (1905).) The idea was strongly supported by H. A. Lorentz, Elektromagnetische verschijnselen in een stelsel dat zich met willekeurige snelheid, kleiner dan die van het licht, beweegt, *Versl. gewone Vergad. wis- en natuurk. Afd. K. Akad. Wet. Amst.* **12**, second part, 986–1009 (1903–4), and by Einstein's special relativity, leading to results deviating from those of Kaufmann and Abraham. The electromagnetic theory of matter had a great ideological influence on scientists because of the fascinating possibility of *explaining matter*. It was shaken and modified by Rutherford's discovery of the nucleus (and the proton) and by Chadwick's discovery of the neutron; which may help to explain why its final overthrow by the theory of nuclear forces was hardly taken notice of.

45. The revolutionary power of special relativity lies in a new point of view which allows the derivation and interpretation of the Lorentz transformations from two simple first principles. The greatness of this revolution can be best gauged by reading Abraham's book (Vol. ii, referred to in note 44 above). This book, which is slightly earlier than Poincaré's and Einstein's papers on relativity, contains a full discussion of the problem situation: of Lorentz's theory of the Michelson experiment, and even of Lorentz's local time. Abraham comes, for example on pp. 143 f. and 370 f., quite close to Einsteinian ideas. It even seems as if Max Abraham was better informed about the problem situation than was Einstein. Yet there is no realization of the revolutionary potentialities of the problem situation; quite the contrary. For Abraham writes in his Preface, dated March 1905: 'The theory of electricity now appears to have entered a state of quieter development.' This shows how hopeless it is even for a great scientist like Abraham to foresee the future development of his science.

46. See H. Minkowski, Space and time, in A. Einstein, H. A. Lorentz, H. Weyl, and H. Minkowski, *The principle of relativity*, Methuen, London (1923) and Dover edn, New York, p. 75. For the quotation from Einstein's letter to Cornelius Lanczos, later in the same paragraph of my text, see C. Lanczos, Rationalism and the physical world, in R. S. Cohen and B. Wartofski (eds), *Boston studies in the philosophy of science*, Vol. 3, pp. 181–98 (1967); see p. 198.

47. See my *Conjectures and refutations*, p. 114 (with footnote 30); also my *Open society*, Vol. ii, p. 20, and the criticism in my *Logic of scientific discovery*, p. 440. I pointed out this criticism in 1950 to P. W. Bridgman, who received it most generously.

48. See A. D. Eddington, *Space time and gravitation*, pp. 162 f., Cambridge University Press (1935). It is interesting in this context that Dirac (on p. 46 of the paper referred to in note 29 above) says that he now doubts whether four-dimensional thinking is a fundamental requirement of physics. (It is a fundamental requirement for driving a motor car.)

49. More precisely, a body falling from infinity with a velocity $v > c/3^{\frac{1}{2}}$ towards the centre of a gravitational field will be constantly decelerated in approaching this centre.

50. See the reference to Troels Eggers Hansen cited in note 27 above; and Peter Havas, Four-dimensional formulations of Newtonian mechanics and their relation to the special and the general theory of relativity, *Revs mod. Phys.* **36**, 938–65 (1964), and Foundation problems in general relativity, in *Delaware seminar in the foundations of physics* (ed. M. Bunge), pp. 124–48 (1967). Of course, the comparison is not trivial: see, for example, pp. 52 f. of E. Wigner's book referred to in note 24 above.

51. See C. Lanczos, *op. cit.*, p. 196.

52. See Leopold Infeld, *Quest*, p. 90. Victor Gollancz, London (1941).

53. See A. Einstein, Die Feldgleichungen der Gravitation, *Sber. Akad. Wiss. Berlin*, part 2, 844–7 (1915); Die Grundlage der allgemeinen Relativitätstheorie, *Annln Phys.*, (4th Ser.) **49**, 769–822 (1916).

53a. I believe that §2 of Einstein's famous paper, Die Grundlage der allgemeinen Relativitätstheorie (see note 53 above; English translation, The foundation of the general theory of relativity, *The principle of relativity*, pp. 111–64; see note 46 above) uses most questionable epistemological arguments *against* Newton's absolute space and *for* a very important theory.

54. Especially Heisenberg and Bohr.

55. Apparently it affected Max Delbrück; see *Perspectives in American history*,

Vol. 2, Harvard University Press (1968), Émigré physicists and the biological revolution, by Donald Fleming, pp. 152–89, especially sections iv and v. (I owe this reference to Professor Mogens Blegvad.)

56. It is clear that a physical theory which does not explain such constants as the electric elementary quantum (or the fine structure constant) is incomplete; to say nothing of the mass spectra of the elementary particles. See my paper, Quantum mechanics without 'the observer', referred to in note 44 above.

I wish to thank Troels Eggers Hansen, The Rev. Michael Sharratt, Dr. Herbert Spengler, and Dr. Martin Wenham for critical comments on this lecture.

Index

Q126.0
P76